MICHAEL FARADAY
AND THE ROYAL INSTITUTION

(The Genius of Man and Place)

MICHAEL FARADAY AND THE ROYAL INSTITUTION

(The Genius of Man and Place)

John Meurig Thomas, FRS

Director,
The Royal Institution of Great Britain and
the Davy Faraday Research Laboratory

Adam Hilger
Bristol, Philadelphia and New York

© IOP Publishing Ltd 1991

British Library Cataloguing in Publication Data

Thomas, John Meurig
  Michael Faraday and the Royal Institution:
  The genius of man and place.
  I. Title
  540.92

  ISBN 0-7503-0145-7

Library of Congress Cataloguing-in-Publication Data are available

Published under the Adam Hilger imprint by IOP Publishing Ltd, Techno House, Redcliffe Way, Bristol, BS1 6NX, England
335 East 45th Street, New York, NY 10017-3483, USA

US Editorial Office:
1411 Walnut Street, Philadelphia, PA 19102

Printed in Great Britain by J W Arrowsmith Ltd, Bristol

For Margaret

MICHAEL FARADAY

1791 Bicentenary Year 1991

Faraday ~ Electricity

THE ROYAL INSTITUTION OF GREAT BRITAIN · W1 · LONDON · 5 MARCH 1991 · LONDON

Lisa a Naomi Thomas

Director's Flat

The Royal Institution,

21, Albemarle St,

London W1X 4BS.

THE ROYAL INSTITUTION
OF GREAT BRITAIN

A. G. Bradbury, 3 Link Road, Leicester.

Contents

Foreword by Sir Brian Pippard, FRS   ix

Preface   Michael Faraday and The Royal
              Institution   xi

Chapter 1   Setting the Scene   1

Chapter 2   Rumford, Davy and The Royal
              Institution   4

Chapter 3   From Errand Boy to the Grand Tour   15

Chapter 4   Faraday's Scientific Contributions   22

Chapter 5   Faraday's Writings   95

Chapter 6   Faraday the Man   116

Chapter 7   Faraday's Influence upon The Royal
              Institution   134

Chapter 8   The Popularization of Science   192

Epilogue   211

Appendix I   Faraday's Discourses & Scientific
              Publications, 1832-1834   212

vii

Contents

Appendix II   Learned Societies to which Faraday
was Elected   215

Appendix III   Faraday's Friday Evening
Discourses, 1835 onwards   217

Appendix IV   Discourses Arranged by Faraday
(before 1862)   220

Appendix V   Part of The Royal Institution
Lecture Calendar, 1933   223

Index   225

Foreword

by
Sir Brian Pippard, FRS, Cavendish Laboratory, Cambridge

The speaker at one of the Royal Institution's Friday Evening Discourses observed that, of all places on Earth, the Institution has seen the greatest number of discoveries per square metre. This must surely be true and, of all the eminent scientists who have worked there, pre-eminent is Michael Faraday, a self-educated man whose inability to master mathematical methods might be thought to rule out a career in the physical sciences. But his genius for experiment, his instinct for the right attack on a problem, his capacity for hard work and, above all, his passion for leaving no loose ends in the solution made of the poor apprentice the most eminent scientist of his time. He early formed the habit (and how rare it is) of recording all his observations as he made them, and adding his immediate thoughts on what they signified. As a result we can probably build up a more complete picture of him than of any other who has changed the course of history to a comparable degree.

Faraday took great trouble to make the latest discoveries of science, his own and those of others, intelligible to the layman, and the tradition he fostered has been kept alive ever since, so that the Royal Institution is as well known for its contributions to education as for its research. The present book is written in the spirit of that tradition, to help meet a need that is greater now than in Faraday's time. The founder, Benjamin Thompson (Count Rumford), aimed to apply science to improve the lives of ordinary citizens in an age when it was perhaps easier than it is now to overlook the risks inseparable from discovery, and to welcome the obvious benefits. Nowadays, however, science has many critics whose views, sometimes ill-founded, command support; and all too often their criticisms pass unanswered, or the answers are lost in the general hubbub. Without determined efforts by scientists to explain what they are doing and why it matters, and to show the sort of people they are, there is a danger that science and technology will cease to attract imaginative young people with something of Michael Faraday's dedication and integrity. If that should happen, we shall all be the poorer, for the quality of civilized life has come to depend on science for its very maintenance, let alone improvement.

Readers of this book will enjoy finding out in how many ways we are indebted to Faraday for his prodigious researches, and will discover why his fellow scientists revered him, and why his many friends loved him. His generosity and singular goodness of heart, as well as his creative intelligence, speak to us across the years and one can never tire of hearing about them again, especially when the tale is told with the understanding and gusto that Professor Thomas has brought to the telling.

Preface

Michael Faraday and The Royal Institution

Having lived and worked for five years in Michael
Faraday's home and laboratory, my initial interest in,
and curiosity about, the great scientist has developed
into a passionate admiration for all that he stood for
and achieved. His scientific and spiritual presence at
the Royal Institution confers a unique aura that
pervades the whole place. One cannot escape it. This
book is meant to share with others my enthusiasm and
admiration for the man, the scientist and the place.

Because of the enormous significance of his work,
the sheer brilliance of its pursuit, execution and
description, and because he was a humble, self-
educated man who conquered and transformed the
world, most people are aware of who Faraday was. But
not everyone knows that Michael Faraday was one of
the most remarkable men of all time. The greatest of
my predecessors as Director of the Royal Institution,
which houses the oldest continuously used scientific
laboratory and the foremost repertory theatre for the
popularization of science in the world, Faraday lived
here for nearly fifty years and worked here for a
somewhat longer period.

In this comparatively slim volume I outline the trajectory of Faraday's exceptional career, analyse and interpret the essence of his discoveries, and examine afresh the wellsprings of his genius. I have endeavoured to do so in such a way as to make the book accessible to all layfolk interested in, but not well-informed about, modern science and especially young people in the arts and in all branches of the sciences who are about to enter tertiary education.

I am greatly indebted to Margaret Gowing, formerly Professor of the History of Science, University of Oxford and Sir Brian Pippard, formerly Cavendish Professor at the University of Cambridge who, having read the first draft, gave me much encouragement, guidance and advice; to Mrs Irena McCabe, Archivist and Librarian at the Royal Institution for invaluable advice and assistance in guiding me to sources on Faraday's life and work here; to Drs Frank James and Brian Bowers for their comments on Chapter 4 and to Mr Peter Branch and Mrs Jean Conisbee for expert help with the illustrations. Miss Sue Horrill has coped magnificently and with equanimity with the conversion of my disjointed original text into a streamlined format. Mrs Maureen Clarke and Mr Al Troyano at Adam Hilger have given exemplary advice on editorial matters, for which I am most grateful. I appreciate the help and support of all these, but of none more so than my wife Margaret who read the entire text and whose concern for clarity and economy greatly improved it. I fondly dedicate this book to her.
Any shortcomings or infelicities that remain however are wholly mine.

Chapter 1

Setting the Scene

Michael Faraday is generally held to be one of the greatest experimental philosophers of all time. According to Albert Einstein he was also responsible along with Clerk Maxwell, for the greatest change in the theoretical basis of physics since Newton. Such was the prodigality of his output and the diversity of his skills that modern chemists, no less than physicists, engineers and materials scientists, regard him as one of the founders of their subjects: some sciences and technologies owe their very existence to his work. No name stands higher in the general esteem of scientists the world over than that of Faraday; and few names are encountered more frequently by first year university students of science than his. He bequeathed to posterity a greater body of pure scientific achievement than any other physical scientist, and the practical consequences of his discoveries have profoundly influenced the nature of civilised life.

Faraday was self-taught: he left school at the age of twelve, and started his career as an errand boy and then as a bookbinder. In none of his four hundred and fifty publications is there a single differential equation, for he knew no mathematics. But calculus, as Clerk Maxwell remarked, is but a part of mathematics, and, according to him, Faraday was 'in reality a mathematician of a very high order — one from whom the mathematicians of the future may derive valuable and fertile methods.'

The story of Faraday's life and work is one of the most romantic and successful in the annals of science; and it is inextricably mingled with the fortunes of the Royal Institution, which he first entered as the result of a chance occurrence in 1812, and where he subsequently lived and worked for nearly fifty years. All his discoveries were made there, initially under the aegis of his brilliant mentor, Humphry Davy; all his educational initiatives, successful to this day, were mounted there.

He worked in the quiet of his basement laboratory. He built his own equipment and he often designed and constructed his own instruments. All but two of his papers were authored solely by him. They are masterpieces of lucidity, self-criticism and insight, and still serve as models for both aspiring and mature scientists. His voluminous notebooks, which detail his ambitions and doubts, successes and failures, and constitute a day-to-day record of the vicissitudes in his experimental work, are likewise a mine of information, of value both to the working scientist and the historian.

In the mid 1820s, Faraday initiated two brilliantly successful educational ventures in the public understanding and popularization of science: the Friday Evening Discourses for lay audiences and the Christmas Lectures for children, both of which still continue at the Royal Institution. Faraday gave the Christmas Lectures on nineteen occasions. His most famous series on *The Chemical History of a Candle*, first published in 1860, has become a classic translated into many languages. (It is still recommended reading in the summer vacation for Japanese schoolchildren.) Faraday himself, through his experiences with young and old audiences at the Royal Institution, became one of the foremost lecturers of his time.

Though of a retiring, almost reclusive nature, and profoundly religious, Faraday was a leading figure in Victorian England. Prince Albert befriended him and among those with whom he interacted were the painters Turner and Constable, the writers Dickens and Ruskin, and the eminent biologists and champions of evolution Charles Darwin and T H Huxley. To these must be added Fox Talbot, one of the founders of photography, Roget, the originator of the thesaurus that bears his name, the polymathic scientists Lord Kelvin and William Wollaston (physician, geologist, chemist), and the philosopher William Whewell, later Master of Trinity College, Cambridge. Several of these gave Friday Evening Discourses at the Royal Institution, which, for much of the period when Faraday was Director there, was the focus of the intellectual, cultural and scientific life of London.

Apart from his towering scientific achievements, Faraday's contributions extended to many other fields. He was an early advocate of science teaching; he advised the National Gallery on the conservation of paintings, the British Museum on the cleaning of the Elgin Marbles and the Corporation of Trinity House on the illumination of lighthouses. He was the founding secretary of the London club, The Athenaeum.

Faraday combined singular gifts of intellectual power, technical virtuosity, intuition and moral perfection. Ben Jonson's remark about Shakespeare applies: 'He was not of an age, but for all time!'

Chapter 2

Rumford, Davy and The Royal Institution

In 1798, the American-born scientist and statesman, Sir Benjamin Thompson, otherwise known as Count Rumford, found himself temporarily jobless. Accused of spying, he fled from America and came to London where he took out British citizenship. Later he became the Elector of Bavaria's principal advisor and head of his military services. He had been sent to London as the Elector's Minister Plenipotentiary and Envoy Extraordinary to the Court of St James. But King George III refused to accept one of his own citizens as a foreign minister. Under the circumstances, Rumford set about devising plans to create the Royal Institution of Great Britain, which, with the support of the President of the Royal Society, Sir Joseph Banks, he founded in 1799

> ... for diffusing and facilitating the general introduction of useful mechanical inventions and improvements and for teaching by courses of philosophical lectures and experiments, the application of science to the common purposes of life.

By 1802, however, Rumford was locked in disagreement with the Managers of the Royal Institution. He disapproved of their plans for future development. And so he left for France and for Anne Lavoisier (the great chemist Antoine Lavoisier's widow whom Rumford later married with disastrous consequences) never to return to London again. Rumford was a colourful character once described[1] as

a loyalist, traitor, spy, cryptographer, opportunist, womaniser, philanthropist, egotistical bore, soldier of fortune, military and technical advisor, inventor, plagiarist, expert on heat (especially fireplaces and ovens) and founder of the world's greatest showplace for the popularization of science, the Royal Institution.

As a scientist he clarified the nature of convection, devised an ingenious instrument for measuring luminosity — his definition of the standard candle was used for a hundred years — and invented a drip coffeemaker. In school textbooks he is best remembered for showing that the drilling of the bore of a cannon supplies apparently inexhaustible quantities of heat, which would be impossible if heat were a material fluid (caloric).

Apart from establishing it, Rumford left his mark on the Royal Institution which, from the outset, was a great success; people flocked to its lectures and exhibitions. In 1801 he had recruited two exceptional Englishmen, Thomas Young (1773-1828), now immortalized through the modulus of elasticity that bears his name, and the Cornishman Humphry Davy (1778-1829). The former was appointed Professor of Natural Philosophy and the latter Assistant Lecturer. Young's scientific accomplishments were formidable. In 1802 he carried out his famous 'fringe' experiment thus resurrecting the wave-theory of light. Later he

5

made seminal contributions to the theory of capillarity and surfaces, as well as devising ingenious optical instruments, writing papers on a variety of medical subjects and facilitating the decipherment of Egyptian hieroglyphics, notably those on the Rosetta Stone. Young's lecturing skills did not, however, match his prowess as a scientist and linguist; his performances before popular audiences were dismal failures.

Figure 1 The inventive Count Rumford, founder of The Royal Institution, standing in front of one of his creations.

Davy, by contrast, was a coruscatingly brilliant lecturer, whose carefully prepared, well-rehearsed, fluently delivered performances and breath-taking demonstrations to lay audiences rapidly became important social functions, and added greatly to the prestige of science and the Institution. He began one of his lectures thus:

The love of knowledge and of intellectual power is a faculty belonging to the human mind in every state of society; and it is one by which it is most justly characterized — one most worthy of being cultivated and extended.

The poet Coleridge said he went to listen to Davy at the Royal Institution 'to renew my stock of metaphors' and claimed 'that had he (Davy) not been the first chemist, he would have been the first poet of the age.' Coleridge went further:

There is an energy, an elasticity in his mind which enables him to seize on and analyse all questions, pushing them to their legitimate consequences. Every subject in Davy's mind has the principle of vitality. Living thoughts spring up like turf under his feet.

Even though initially Rumford felt ambiguous about them, there is no doubt that Davy's efforts ultimately met with his approval, even from afar. Before his appointment by Rumford to the Royal Institution, Davy had demonstrated how the gas nitrous oxide, otherwise known as 'laughing gas', could be used as an anaesthetic. He had also convinced himself, on reading Alessandro Volta's (1745-1827) famous letter to Sir Joseph Banks on 'A continuous source of electricity by the mere contact of two dissimilar metals', that chemical action was responsible for the production of electricity. Davy argued that the converse could, therefore, be true: certain chemicals could be produced using electricity.

Figure 2   Thomas Young, physician, physiologist, physicist, philologist, appointed Professor of Natural Philosophy by Rumford. According to Helmholtz 'he excited the wonder of his contemporaries who, however, were unable to rise to the heights at which his daring intellect was accustomed to soar.'

This in turn led to his discovery of sodium, potassium, calcium, barium, strontium and magnesium. Later he isolated boron and (in France) clarified the nature of iodine. Davy's vitality, vision and ability soon established the laboratory in the basement of the Institution as among the finest and best equipped in the world. (It predated the Clarendon and Cavendish laboratories of the Universities of Oxford and Cambridge by more than sixty years.) When money was scarce for establishing new facilities,

Davy appealed to enlightened subscribers, using words that nowadays are beloved by professional fund-raisers, as his plea for funds to install the world's most powerful voltaic battery (in excess of 5,000 volts at high current) testifies, see figure 4.

Figure 3 Sir Humphry Davy, poet, innovator, man of action. He combined elegance of literary expression with brilliant scientific discovery.

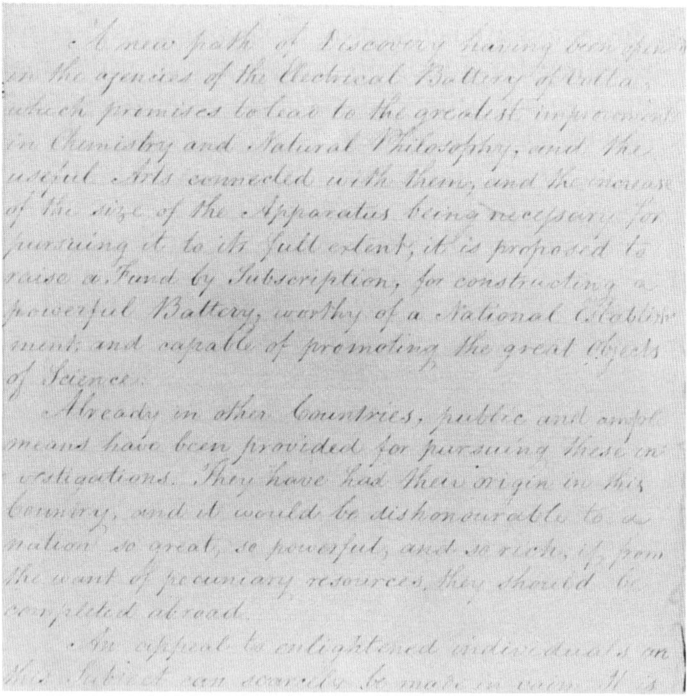

A new path of Discovery having been opened in the agencies of the Electrical Battery of Volta, which promises to lead to the greatest improvements in Chemistry and Natural Philosophy, and the useful Arts connected with them; and the increase of the size of the Apparatus being necessary for pursuing it to its full extent, it is proposed to raise a Fund by Subscription, for constructing a powerful Battery, worthy of a National Establishment and capable of promoting the great Objects of Science.

Already in other Countries, public and ample means have been provided for pursuing these investigations. They have had their origin in this Country, and it would be dishonourable to a nation so great, so powerful, and so rich, if from the want of pecuniary resources, they should be completed abroad.

An appeal to enlightened individuals on this subject can scarcely be made in vain. It is

Figure 4 Excerpt from the Report of Managers, 1808, describing Davy's 'appeal to enlightened individuals'.

Davy left for posterity a number of revealing insights into his character and personality. In *The Collected Works of Sir Humphry Davy, Bart*, edited by John Davy, 1839-40, we read about him as a teenager:

> *After reading a few books, I was seized with a desire to narrate, to gratify the passions of my youthful auditors. I gradually began to invent, and form stories of my own. Perhaps this passion has produced all my originality, I never had a memory. I never loved to imitate, but always to invent: this has been the case in all the sciences I have studied. Hence many of my errors.*

Early in the summer of 1801, just three months after his arrival at the Royal Institution, Davy wrote to his friend in Bristol, John King:

> *The voice of fame is still murmuring in my ears — my mind has been excited by the unexpected plaudits of the multitude — I dream of greatness and utility — I dream of science restoring to nature what luxury, what civilization have stolen from her — pure hearts, the forms of angels, bosoms beautiful and panting with joy and hope — My labours are finished for the season as to public experimenting and public enunciations. My last lecture was on Saturday evening. Nearly 500 persons attended — ... There was respirations of nitrous oxide: and unbounded applause. Amen. Tomorrow a party of philosophers meet at the Institution to inhale the joy inspiring gas — It has produced a great sensation. Çá ira ... I have been nobly treated by the Managers. God bless us. I am about 1,000,000 times as much a being of my own volition as at Bristol. My time is too much at my own disposal. So much for egotism — for weak, glorious, pitiful, sublime conceited egotism.*

For twelve years Davy was employed full-time at the Royal Institution, becoming its Professor of Chemistry in 1802 and its Director in 1804. Thereafter he held the post of Honorary Professor up to his death in 1829. The roll-call of his achievements, by any standards, is extraordinary. In addition to his discovery of the elements listed above, he invented the electric arc, the miners' safety lamp, methods for bleaching cloth, copying paintings on ceramics, tanning leather and arresting the corrosion of ships through cathodic protection (a technique which he also invented). By establishing the correct composition of hydrochloric acid he disproved Lavoisier's assertion that all acids contain oxygen. He made pioneering contributions to geology, mineralogy and agricultural chemistry which

11

led to the publication, in 1813, of his *Elements of Agricultural Chemistry*. His Bakerian Lecture of 1806 to the Royal Society 'On Some Chemical Agencies of Electricity' described the work which earned him the Napoleon Prize from the Institut de France (1807). In his period as President of the Royal Society (1820-27), during which Thomas Young was its Foreign Secretary, Davy helped to establish the Athenaeum Club, the London Zoological Society (and the London Zoo in Regent's Park) and the Geological Society. Throughout his life he cultivated his artistic propensities. Some of his earliest works were poems. He was a central figure in Wordsworth's circle of friends (which included Southey, Coleridge and Walter Scott) and frequently entertained them at the Royal Institution. Davy's poetry, as well as his chemistry, is mentioned by George Eliot in *Middlemarch*. The poetic flair that Davy brought to his science is beautifully illustrated in the opening paragraph of his article on 'Some experiments and observations on the colours used in painting by the Ancients', published in *Phil. Trans. Roy. Soc.*, 1815. (This is one of the first examples of the application of science to archaeology and of the scientific analysis of pigments.)

The importance the Greeks attached to pictures, the estimation in which their great painters were held, the high prices paid for their most celebrated productions, and the emulation existing between different states with regard to the possession of them, prove that painting was one of the arts most cultivated in ancient Greece; the mutilated remains of the Greek statues, notwithstanding the efforts of modern artists during three centuries of civilization, are still contemplated as the models of perfection in sculpture, and we have no reason for supposing an inferior degree of excellence in the sister art, amongst a people to whom genius and taste were a

12

kind of birthright, and who possessed a perception,
which seemed almost instinctive, of the dignified, the
beautiful, and the sublime.
   The works of the great masters of Greece are
unfortunately entirely lost.   They disappeared from
their native country during the wars waged by the
Romans with the successors of Alexander, and the later
Greek republics; and were destroyed either by accident,
by time, or by barbarian conquerors at the period of the
decline and fall of the Roman Empire.

Davy's circle of friends was remarkable. A trusted
adviser of the highest society, he was welcomed as
honoured guest at the great country houses.   A
painting that now hangs in the Tate Gallery shows
Davy present, along with other members of the
illuminati, at Mr Coke's annual sheep-shearing at
Halkham. He stayed with the Duke of Bedford at
Woburn, with Lord Sheffield in Sussex, and with Lord
Byron in Ravenna.   The Rev Sydney Smith, that
resolute champion of parliamentary reform, who filled
the lecture theatre of the Institution to overflowing
when he spoke on Moral Philosophy in 1804, was a
member of Davy's circle. So was the intellectually
omnivorous William Hyde Wollaston, a leader in
mineralogy, botany and chemistry, and founder of
powder metallurgy and much else.   Wollaston's
method of producing sheets and wires of platinum,
and his discovery of the noble metals palladium and
rhodium were to influence the course of physical
science profoundly throughout the 19th and much of
the 20th century. They made possible a good deal of the
early work at the Royal Institution, by Davy and his
successor especially, in electrochemistry and   the
catalysed combustion of gases.

One of Humphry Davy's last acts as Director was to
interview a young man named Michael Faraday.

Figure 5 Library of the Royal Institution in 1809, from a drawing by Rowlandson in *The Microcosm of London.*

Notes

1. W H Brock, *New Scientist*, 27 March 1980.

Chapter 3

From Errand Boy to the Grand Tour

Michael Faraday, the third son of a journeyman blacksmith, was born on 22 September 1791 in Newington Butts on the outskirts of London. His early education was extremely simple, consisting of little more than the rudiments of reading, writing and arithmetic. At the age of thirteen he became an errand boy to Mr Riebau, a kindly French émigré bookbinder and bookseller, who later took him on as an apprentice. In the years that followed he read the books that he was given to bind. One of them was a copy of the *Encyclopaedia Britannica*, 3rd edition, 1797, where he came across the article by the wayward chemist, Mr James Tytler on electricity, a topic which captured his interest. He also bound and read *Conversations on Chemistry* by Mrs Marcet, the wife of a Swiss doctor, who had published her book in 1809 for audiences created by Humphry Davy.

Guided by the advice on self-improvement contained in another book that came his way, *The Improvement of the Mind* by the famous hymn writer Isaac Watts, he kept a 'common place book' or notebook in which he jotted down ideas, facts, quotations and questions as they occurred to him. The earliest one has the following description on the first page:

> *A Collection of Notices, Occurrences, Events etc., relating to the Arts and Sciences collected from the Public Papers, Reviews, Magazines and other miscellaneous works. Intended to promote both Amusement and Instruction and also to corroborate or invalidate those theories which are continually starting in the world of science.*

We learn from this 'common place book' that Faraday was prompted to carry out simple experiments in chemistry, encouraged by his kindly employer. He soon acquired, through his own efforts, enough expertise to decompose magnesium sulphate using a voltaic pile, the construction of which he described in a letter to his friend Benjamin Abbott:

> *I, Sir, I my own self, cut out seven discs (of zinc) the size of halfpennies each. I, Sir, covered them with seven halfpences and interposed between seven or six pieces of paper soaked in a solution of muriate of soda.*

In 1812 a Mr Dance, one of the customers at Riebau's shop, gave Faraday a ticket to attend the last four lectures of a course given by Davy at the Royal Institution (figure 6). Faraday was enthralled by Davy's performance. He took copious notes, re-wrote them, added illustrations and an index, and bound them with his own hand (figure 7). He later revealed that, at that time,

> *My desire to escape from trade, which I thought vicious and selfish, and to enter into the services of Science, which I imagined made its pursuers amiable and liberal, induced me at last to take the bold and simple step of writing to Sir H. Davy expressing my wishes and a hope that, if an opportunity came in his way, he would favour*

my views: at the same time I sent the notes I had taken of his lectures.

Figure 6 Plaque commemorating Faraday's first entry to The Royal Institution.

Davy, impressed with Faraday's confidence and zeal, interviewed him early in 1813. There was no vacancy at the Institution and Davy advised him to stick to his craft. 'Science', said Davy, 'is a harsh mistress.' In February of that year, however, an assistant in the Laboratory was dismissed for brawling and Davy offered Faraday the post at a salary of twenty-five shillings a week with two rooms at the top of the Institution in 21, Albemarle Street.

Faraday began work as Davy's assistant on 1 March 1813. Within a few days he was helping with research; and within a few weeks Davy entrusted him with the preparation of samples of the newly discovered nitrogen trichloride. A little later Davy and Faraday suffered as a result of the premature explosion of this capricious substance. Faraday also provided skilful preparative assistance at lecture experiments, a task that induced him to ruminate on the art of lecturing. In his notebook he wrote:

17

A lecturer should appear easy and collected, undaunted and unconcerned, his thoughts about him and his mind clear for the contemplation and description of his subject. His action should be slow, easy and natural consisting principally in changes of the posture of the body, in order to avoid the air of stiffness or sameness that would be otherwise unavoidable.

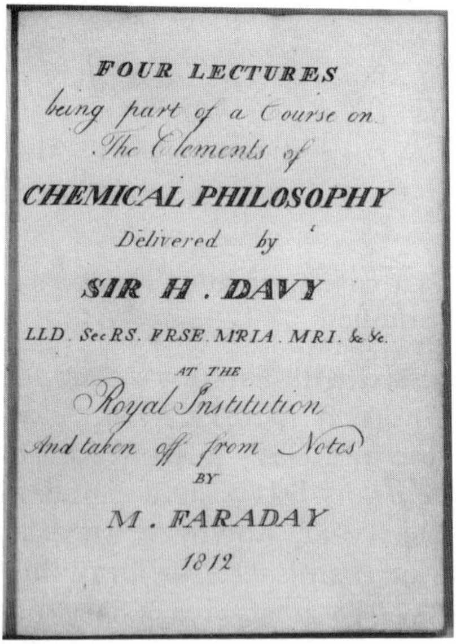

FOUR LECTURES
being part of a Course on
The Elements of
CHEMICAL PHILOSOPHY
Delivered by
SIR H . DAVY
LLD. Sec RS. FRSE. MRIA. MRI. &c &c.
AT THE
Royal Institution
And taken off from Notes
BY
M . FARADAY
1812

Figure 7 Title page of Faraday's notes on Davy's lectures, 1812.

Davy's expertise, panache and general celerity of action greatly facilitated Faraday's progress as an experimentalist. The laboratory and environment in which he worked was, in modern parlance, 'a centre of excellence'. If it is true that, as Rutherford once remarked, a centre of excellence is, by definition, a place where second class people may perform first class work, how much more could be expected from Michael

Faraday, endowed as he was with such extraordinary native talent and guided by Humphry Davy, one of the brightest scientific stars in the European firmament?

Further good fortune was soon to come Faraday's way. It so happened that Davy had planned to embark on an extended tour of Europe with his wife in the autumn of 1813. Even though England and France were then at war, Davy's reputation and Napoleon's regard for science were such that Davy was able to proceed with his plans. He invited Faraday to accompany the party as his secretary and scientific assistant. On 13th October of that year, they set off from Plymouth for France, Italy and Switzerland carrying the requisite scientific equipment for the experiments to be undertaken *en route*. In Paris, they experimented with the newly discovered element, iodine, with which Ampére provided them, and naturally, in view of their experience with nitrogen trichloride, prepared the explosive nitrogen triiodide; they attended a lecture by Gay-Lussac and, before leaving that city, met other savants such as Arago, Humboldt and Cuvier, and dined with Count Rumford whose marriage had collapsed five years earlier (Rumford died, a broken man, in 1814).

In Genoa, they studied the electrical discharge from the torpedo fish, and spent some time in Florence where, with the aid of the great lens in the Accademia del Cimento, they conducted 'the grand experiment of burning the diamond', and proved that the latter like graphite was none other than pure carbon (a view which, several decades later, was still disputed by many including James Dewar, Director of the Royal Institution a century later). Davy's idea of using the sun's rays focused on the diamond was that it permitted clean combustion of the gemstone to take place within a closed system. The only products were the oxides of carbon.

They visited Vesuvius, made the acquaintance of Volta in Milan, and Faraday wrote notes on the firefly and the glow worm. They collected the natural flammable gas at Pietra Mala, and identified it as methane in the laboratory of the Accademia in Florence. Their visit to Italy afforded opportunities to examine papyri and pigments from ancient sites, opportunities from which Davy, as we saw earlier (page 12), extracted full scientific advantage.

Faraday witnessed the commanding position which Davy's gifts and reputation as a natural philosopher had acquired for him in European science; he also, sadly, had occasion to realise that the homage deservedly paid to his great master had led, perhaps, to some deterioration in those personal qualities which might be expected to accompany intellectual eminence. Davy's pre-eminence had made him vain and inconsiderate.

The European tour brought Faraday into contact with Professor Gustave de la Rive of Geneva and his son Auguste, with each of whom he corresponded over the ensuing decades (see Chapter 5). For Faraday the extended visit, with Europe's premier scientist as his travelling tutor, and constant opportunities to acquire a working knowledge of French and Italian constituted, in effect, his university education.

In May 1815, Faraday began work again at the Royal Institution, and shortly thereafter took possession of living apartments in the building. His official post was 'Assistant in the Laboratory and Mineral Collection and Superintendent of the Apparatus.' A few months later, the Institution increased his salary to £100 per annum, where it remained until 1853.

Figure 8 This site in Albemarle Street, off Piccadilly, has been occupied since the Royal Institution was founded in 1799. The classical exterior was added to the building by Villiamy in 1838. (From a watercolour drawing by T H Shepherd.)

Chapter 4

Faraday's Scientific Contributions

Largely because of their immediacy in the world of technology, and also because of their brilliant quantitative interpretation and extension by Clerk Maxwell, Faraday's discoveries in electricity and magnetism tend to dwarf, at least in the mind of the modern physicist, engineer or cosmologist, his contributions to other branches of science. But, as we have indicated earlier, he is also one of the founders of organic and analytical chemistry, of electro- and magneto-chemistry and, indeed, of most sub-disciplines of the physical sciences.

To some extent it is artificial to distinguish his work as a chemist from that as a physicist. (He himself disliked the term physicist, and preferred to be described as a natural philosopher.) It is, however, convenient to enumerate his contributions under two, somewhat arbitrary headings: Chemistry on the one hand, and Electricity and Magnetism on the other (figures 9 and 10).[1] These enumerations reflect the vast extent of Faraday's canvas and the exquisite touch of his brush.

1816	(With Davy) Evolution of Miners' **Safety Lamp**.
1818-24	Preparation and properties of **alloy steels** (study of Indian Wootz). **Metallography**.
1812-30	**Analytical chemistry.** Determination of purity and composition of: clays, native lime, water, gunpowder, rust, dried fish, various gases, liquids and solids.
1820-26	**Organic chemistry.** Discovery of: benzene, iso-butene, tetrachloro-ethene, hexachlorobenzene, isomers of alkenes and of napthalene sulphonic acids (α and β), **vulcanization** of rubber. **Photochemical** preparations.
1825-31	Improvements in the **production of** optical grade **glass**.
1823, 1845	**Liquefaction** of gases (H_2S, SO_2 and six other gases). Recognized existence of **critical temperature** and established reality of **continuity of state**.
1833-36	Electrochemistry and the electrical properties of matter. **Laws of electrolysis.** **Equivalence** of voltaic, static, thermal and animal electricity. First example of **thermistor** action. **Fused salt** electrolytes; **superionic conductors**.
1834	**Heterogeneous catalysis:** **Poisoning** and inhibition of surface reactions. **Selective adsorption;** wettability of solids.
1835	**'Plasma' chemistry** (discharge of electricity through gases).
1836	**Dielectric** constant, **permittivity**.
1845-50	**Magneto-chemistry** and the magnetic properties of matter. **Magneto-optics. Faraday effect.** Diamagnetism. Paramagnetism. Anisotropy.
1857	**Colloidal metals. Scattering of light.** Sols and hydrogels.

Figure 9 Faraday's principal contributions to chemical science.

23

1821	Electromagnetic rotations.
1831	Electromagnetic induction.
	Acoustic vibrations.
1832	Identity of electricities from various sources.
1833	Electrolytic decompositions.
1835	Discharge of electricity through evacuated gases. (Plasma physics and chemistry.)
1836	Electrostatics. Faraday cage.
1845	Relationship between light, electricity and magnetism; diamagnetism; paramagnetism.
1846	'Thoughts on ray vibrations.'
1849	Gravity and electricity.
1857	Time and magnetism.
1862	Influence of a magnetic field on the spectral lines of sodium.

Lines of force and the **concept of a field.** The energy of a magnet lies outside its perimeter. The notion that light and magnetism and electricity were interconnected.

Figure 10 Faraday's principal contributions to physical science.

To appreciate Faraday's meteoric rise as a natural philosopher and the astonishing breadth and depth of his output, it is instructive to sketch, chronologically, both the milestones in his career as well as his specific achievements and their consequential impact, starting from the date of his re-appointment in the Royal Institution. It is also valuable to include in this sketch other incidents and events not of a scientific nature which helped to mould Faraday's character.

In 1816, he helped Davy with the invention of the miner's safety lamp[2]; his **first paper** appeared that year ('Analysis of the native caustic lime from Tuscany'); and he gave his first lecture on some of the chemical elements (to the City Philosophical Society, many of whose members later migrated to the Society of Arts of which he became a member). He was also recruited by W T Brande to help with editorial duties pertaining to the *Quarterly Journal of Science*, published at the Royal Institution. (W T Brande, 1788-1866, was appointed Professor on Davy's departure as Director in 1812. Prior to that he had been lecturing to medical students in Windmill Street. He was a rather ordinary scientist, but well organized; he retained his post at the Institution for over forty years, his main task being to teach chemistry at 9 am every morning in the basement laboratory.)

By 1819 Faraday was already the foremost **analytical chemist** in Britain, specializing on water, clays and various alloys. He was in demand as an expert witness in various litigations; and he had started with James Stodart, a surgical instrument maker, his pioneering work on the **composition and preparation of alloy steels** which led him into correspondence with Josiah John Guest of Dowlais, South Wales, a world centre for steel making at that time. More than a century later Sir Robert Hadfield, FRS, after examining many of the samples that Faraday had prepared, claimed that he:

> .... *was undoubtedly the pioneer of research on special alloy steels; and had he not been so much in advance of his time in regard to general metallurgical knowledge and industrial requirements his work would almost certainly have led immediately to practical development.*

Some of the 'rustless' platinum-containing steels that he and his collaborator James Stodart fashioned into razors and gave to their friends in the 1820s are still extant.

Clays

No 1.   Cornwall Clay — dried
        Silix   53.6
        Alumini   45.6
        Iron oxide   .4

                        99.6

No 2.   Flintshire Clay — dried
        Silix   59.3
        Alumini   40.0
        Iron oxide   .3

                        99.6

Mr. Faraday has to apologise for the delay of these analyses but Workmen in the Laboratory have retarded the usual operations there.

Royal Institution
Feby. 12th 1819

Figure 11 Excerpt of a letter now in the Wedgwood Museum, Stoke-on-Trent, describing Faraday's analysis of clays. (By kind permission of the Trustees.)

In 1820 he **discovered** and **established** the chemical **formula** of two **new compounds** of carbon — tetrachloroethene and hexachloroethane — and a new compound of carbon, iodine and hydrogen. The first of these is like ethene (more commonly known as ethylene) in which the four atoms of hydrogen are replaced by chlorine. (Tetrachloroethene is now used extensively as cleaning fluid. It is also the material used as the 'detector' in the underground solar neutrino experiment in the United States.) The second is related to a constituent of natural gas known as ethane but in which the six atoms of hydrogen are replaced by chlorine. The respective chemical formulae are C_2Cl_4 and C_2Cl_6. Unlike his contemporaries, Faraday did not invoke the notion of 'vital forces' that was thought at that time (by the great Swedish scientist Berzelius especially, even up to 1849) to govern the behaviour of organic compounds. Arguments about vitalism abounded whenever compounds containing carbon were discussed. In this phase of his work Faraday used sunlight as a means of effecting chemical conversions. This is one of the earliest examples of preparative photochemistry, now an important technique in organic chemistry.

Figure 12 A model reconstruction of the chemical laboratory in the basement of the Royal Institution as it was when Faraday worked there.

It was also in 1820 that he became engaged to the daughter of a silversmith, Sarah Barnard, whom he married within a year. She was the sister of someone he had met at the City Philosophical Society. Faraday's parents and those of Sarah Barnard were members of the Sandemanian Church, a small strict body of what would now be looked upon as 'literalist' belief. Like the Quakers, the ascetic Sandemanians believed in lay clergy and were opposed to the accumulation of wealth. Although he did not formally join the Sandemanians until July 1821, Faraday had been regularly attending Sunday morning church meetings ever since he was a child.

Figure 13 The young Faraday, approximately 30 years of age, at his bench.

The First Electric Motor

After reading Ørsted's famous paper in 1820 on how a compass needle is affected when brought close to a wire carrying an electric current, and after writing a definitive historical survey of what was then known about the connection between magnetism and electricity up to 1821, Faraday demonstrated, in September that year, that a wire carrying an electric current rotated around a stationary magnet. His survey[3] was itself no mean task. Before him came the discovery of many fundamental phenomena such as electric and magnetic attractions and repulsions, and the electric current and its effects. Then came Coulomb and Poisson (and Cavendish who did not publish his great discoveries), who, by following the path pointed out by Newton, and making the forces which act between bodies the principal object of their study, founded the mathematical theories of electric and magnetic forces. Next came Ørsted's cardinal discovery (figure 14), followed by Ampère's whereby two wires carrying currents also exert forces on each other (figure 15). Thus the field of electro-magnetic science was already very large when Faraday first entered it. His discovery (figure 16) was sensational, and it soon earned him international fame.

This so-called **electromagnetic rotation**, which enabled him to devise a **primitive electric motor**, could also occur in the earth's magnetic field, a fact that he demonstrated excitedly to his wife on 25 December 1821.

The publicity given to Faraday, and the unfortunate, but unjustified, feeling that he had purloined the idea of electromagnetic rotation from Wollaston, led to

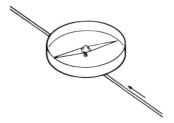

Figure 14 Ørsted discovered that the needle of a compass was deflected by a current-carrying wire. This stimulated Faraday's work in electromagnetism.

ill-feeling between him and Davy. Faraday had been present when Wollaston and Davy had earlier discussed how best to achieve electromagnetic rotation. But he pursued his own line of thinking and arrived at a solution quite distinct from the one favoured by Wollaston. Faraday felt especially grieved when Davy, a close friend of Wollaston's, insinuated at a public discussion at the Royal Society that Faraday's breakthrough had, in truth, originated with Wollaston. Davy later claimed he had been misreported; and Faraday later recognized that he should have shown Wollaston his paper prior to publication. He had indeed endeavoured to do so, but Wollaston was 'out of town', and Faraday's urge to publish was, as ever, strong. However, he openly acknowledged Wollaston's contribution, thereby assuaging the latter's suppressed anger.

[Hans Christian Ørsted (1777-1851) visited Faraday in 1822. They had much in common, especially their interest in electromagnetism and chemistry.[4] Ørsted's first notable achievement scholastically was a thesis on Immanuel Kant for the University of Copenhagen, where he later became a member of the faculty of

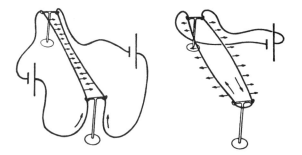

Figure 15 Ampère discovered that wires carrying an electric current exert forces on each other.

medicine. His friendship with Faraday influenced him profoundly. On his return to Copenhagen he founded the Danish Society for the Promotion of Scientific Knowledge, modelled on the Royal Institution. In 1825 he tackled and conquered a problem that had defeated Humphry Davy: the isolation of elemental aluminium. (It is interesting to recall that Davy's spelling — aluminum — is now used in the U S and Canada; elsewhere it is spelled aluminium.) The technique of electrolysis which had failed Davy in his attempt to isolate aluminium in the early 1800s proved successful in the hands of a twenty-two year old Oberlin College (Ohio) student, Charles Martin Hall, in 1886. Hall isolated metallic aluminium by electrolysis of a fused salt (see page 51) by extending work first carried out by Faraday in 1833. Ørsted's method of isolation was chemical and it entailed the use of potassium amalgam, first prepared by Davy.]

In 1823 Faraday discovered and analysed the **first recorded example of a gas hydrate**, a material now termed **a clathrate** because the guest molecule, chlorine in this case, is buried (*clathros*, to bury, in Greek) inside

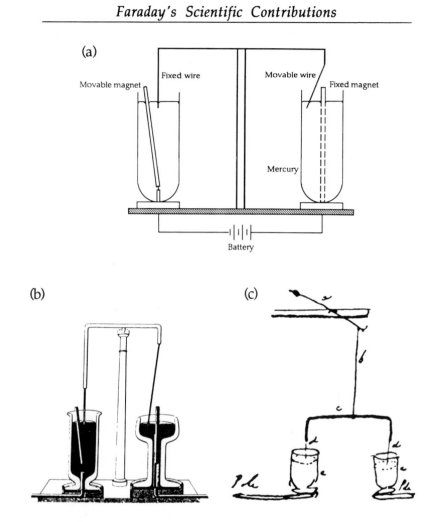

Figure 16 (a) Schematic illustration of Faraday's 'electromagnetic rotations' apparatus with which he showed that a wire carrying a current could be made to rotate around a stationary magnet, and a magnet to rotate around a stationary wire. (b) Published version of the apparatus made by Newman to Faraday's instructions. (c) Faraday's sketch of the apparatus from his diary entry of 22 December 1821.

the accommodating host, the crystallized water. That year also, Faraday **liquefied chlorine**, an achievement that aroused the jealousy of Davy who considered that he had initiated the work and was entitled to the credit. During 1823, and part of 1845 when he returned again to studying the **liquefaction of gases**, Faraday succeeded in liquefying ammonia, carbon dioxide, sulphur dioxide, nitrous oxide, hydrogen chloride, hydrogen sulphide, cyanogen, and ethene (ethylene). He also became the first to recognize the **existence of the critical temperature**, above which, no matter how high the pressure, liquefaction of a gas will not ensue. This also disproved the erroneous and rigid view, then in vogue, that substances fall into three firm categories: solids, liquids and gases. Faraday had demonstrated the **continuity of matter**: the three states were convertible from one to the other.

In 1824, even though Humphry Davy, by then its President, opposed the nomination, the Royal Society elected him a Fellow. The reasons for Davy's action are not entirely clear, but envy undoubtedly contributed to it. In February of that year, Faraday was appointed secretary (its first) of the newly founded London club, The Athenaeum, established largely through Davy's and J W Crocker's (First Secretary of the Admiralty) efforts 'for literary and scientific men and followers of the fine arts'. It is quite possible that he was dragooned into this post by Davy. Prospective members were invited to write to Mr Faraday at the Royal Institution, where the first meetings of the club were held. Faraday relinquished his secretaryship after three months, doubtless because it interfered with his other work. As Davy made progressively fewer appearances, owing to the demands of Presidency of the Royal Society and to his failing health, a new Pharaoh arose in the Royal

Institution. Faraday had begun his lectures to members, and his modesty and charm soon won their hearts.

The Discovery of Benzene

In 1825, Faraday became Director of the Laboratory under the superintendence of the Professor of Chemistry, W T Brande (see page 25). This was the year when he discovered the important organic substance 'bicarburet of hydrogen', later named benzene. His interest in the work that led to this discovery had been aroused by the fact that at the bottom of containers of gas delivered to the Royal Institution by his brother Robert, who worked for the London Gas Company, was a clear aromatic liquid. This turned out to be benzene, which Faraday soon produced by an independent method involving the thermal treatment of fish oil. The consummate skill with which he conducted its characterization and established its chemical formula elicited the admiration of Berzelius in Stockholm. (This was achieved by the careful quantitative analysis of the amounts of carbon dioxide and water vapour produced when the benzene is completely combusted in oxygen. The investigation also demanded the use of effective methods of purification, for which purpose Faraday used fractional distillation and re-crystallization. So skilled was Faraday that the melting and boiling point reported by Faraday for the bicarburet of hydrogen are very close to currently accepted values.) Shortly thereafter he **discovered isobutene** (isobutylene). Noticing that its empirical formula (CH_2) was the same, but its properties quite distinct from that of ethene, he recognized that he had encountered molecular isomerism, though it was not thus named at the time. He later **established** the **formula** of **naphthalene**, and **prepared** two crystalline **sulphonate derivatives** of this substance.

The Chemical and Structural Formulae of Substances Discovered by Faraday

Substance	Chemical Formula	Structural Formula
C_2Cl_4	Tetrachloroethylene (Tetrachloroethene)	
C_2Cl_6	Hexachloroethane	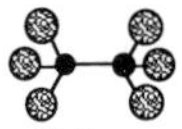
$(CH_2)_4$	Isobutylene (Isobutene)	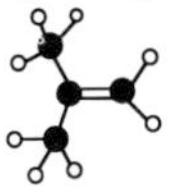
C_6H_6	Bicarburet of Hydrogen (Benzene)	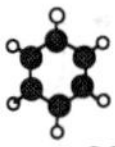
$C_{10}H_7SO_3H$	Naphthalene Sulphonic Acid	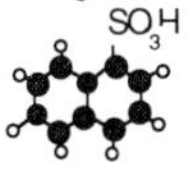

Figure 17 The chemical formulae of the organic compounds discovered or first properly identified by Faraday, along with their subsequently determined structural formulae.

All this was pioneering work that led the way to the aniline dye industry and to other sectors of the chemical and explosive industry. Benzene and naphthalene are now known to be planar molecules, the premier members of an enormous family of compounds called aromatic hydrocarbons. Such molecules, benzene especially, apart from their roles as fuels, are important building blocks in the modern pharmaceutical industry. They are now used in the synthesis of many useful compounds.

Also in 1825, the government chose him through a Royal Society committee chaired by its President, Humphry Davy, to spearhead a project, which lasted until 1830, to **improve the optical quality of glass** (for telescopes). This investigation brought him into regular contact with Thomas Young and J W H Herschel, the versatile son of the famous astronomer father (Sir William Herschel) who with Davy were key members of the committee responsible for overseeing his work. This work was the subject of Faraday's (first) Bakerian Lecture to the Royal Society, given in three parts on 19 November, 3 December and 10 December 1829.

Figure 18 Michael Faraday from the drawing (1852) by George Richmond.

Only a relatively narrow line divides altruistic patronage on the one hand and insidious exploitation on the other. This remark springs to mind when considering Davy's dealings with Faraday in the last decade of Davy's life. Although Davy had relinquished his permanent post at the Royal Institution as early as 1812, his Honorary appointment meant that he could still exert influence there, much of it to Faraday's good. But Davy's attitude to Faraday became much less avuncular and faintly dictatorial and after their relationship had cooled, Faraday grew increasingly aware that much of the work that came his way was at the instigation, even the command, of Davy. From social undertakings, such as the secretaryship of The Athenaeum which Davy had urged him to undertake, Faraday could, with impunity, resign. But he could not easily disengage himself from the tedious, extremely time-consuming endeavours required for the project on optical quality glass. Faraday had wanted to explore new pastures, but this was not possible as long as he had to labour under the supervisory eye of the Royal Society committee that had initiated this project. The death, in 1829, of Humphry Davy and Thomas Young, must have relieved Faraday's growing anxiety about his future.

In 1826 Faraday inaugurated two eminently successful educational ventures in the **popularization and public understanding of science**: the **Christmas Lectures** for young children ('juvenile auditory'), and the **Evening Discourses** to members of the Royal Institution and their guests. The subject of the first series of Christmas Lectures was 'Astronomy', presented by Faraday's friend, J Wallis. Details of the first five Friday evening meetings initiated in 1826 at which Faraday discoursed were as follows:

3 February: 'Caoutchouc'[5];

10 February: 'Mr Brunel's work on liquefied gases as mechanical agents';

3 March: 'Lithography';

7 April: 'All solid and fluid bodies give off or are surrounded by a vapour of their own whatever the temperature';

5 May: 'Singular power of hydrocarbon of entering into union with sulphuric acid'.

Friday Evening Discourses with a similar wide variety of subject matter continue to this day (see Chapters 7 and 8). In addition to all the preparation required for these exercises in the public presentation of science, Faraday published sixteen articles that year, including one bearing the interesting and useful title: 'Plan of an extended and practical course of lectures and demonstrations on chemistry', delivered in the laboratory of the Royal Institution by W T Brande and Michael Faraday.

In 1827 Faraday **declined** the **invitation** to become the first **Professor of Chemistry** at **University College, London**. But in 1829 he was **appointed Professor of Chemistry** at the **Royal Military Academy** at Woolwich, a post which entailed twenty-five lectures a year, and which he continued for some twenty years.[6] This took him to Woolwich for at least one day, sometimes two, each week during term time. That year saw the appearance of the first edition of his 646-page book on *Chemical Manipulation*, a monograph of which the Nobel Laureate, Sir Robert Robinson, the premier chemist of his era, could say in 1931:

This is a treatise on methods used in chemical work, all of which had been proved and many devised by himself, a book which may be profitably studied by the chemical student of today. For although he was limited by the want of the conveniences of a modern chemical laboratory, such as Bunsen burners, Liebig condensers, rubber tubing in lengths, modern wash-bottles and glass taps, the work indicates the source of much of present-day practice and recalls a time when the chemist had also to be the craftsman. ... He recommends good general illumination for a laboratory, he also makes a point of admitting at some place direct sunlight, as 'the solar rays have been found highly influential in causing chemical change.' A striking example is obtained when he gets little chemical action between the vapour of his benzene and chlorine until he subjects the mixture to sunlight, a type of reaction which forms the subject matter of photochemistry.

Over the Christmas-New Year period, 1829-30, Faraday gave the Christmas Lectures for children on the topic of 'Electricity'. And during 1830 he gave six Friday Evening Discourses on topics as varied as the transmission of musical sounds through solids, the application of new principles for the construction of musical instruments (both based on Wheatstone's discoveries), a geological subject and the oxidation of iron.

[Charles Wheatstone (1802-1875) made many inventions in the fields of acoustics, electric telegraphy, optics and music. He started his career as a musical instrument maker in London and he built the first concertina, a type of small accordian. He was appointed Professor of Experimental Philosophy in King's College, London in 1834, the same year in which he used a revolving mirror in an experiment to measure the speed of electricity in a conducting wire — a principle that was later used to determine the velocity of light. With W F Cooke he patented an early telegraph in 1837; the year he began his work on submarine telegraphy in Swansea Bay. He invented the kaleidoscope and the stereoscope, a device for observing pictures in three dimensions, now widely used in viewing X-ray, electron microscopic and aerial photographs. But Wheatstone's Bridge, an instrument for measuring electrical resistance, was not invented by him but by the mathematician Christie: it was Wheatstone who popularized it.]

Discovery of Electromagnetic Induction: the First Transformer and Dynamo

Faraday's most momentous discovery, that of electromagnetic induction, which soon brought forth the electric generator (i.e. the dynamo) and the transformer, was made on **29 August 1831**, the date now acknowledged as **the birth of the electrical industry.** Ever since he had read Ørsted's epoch-making paper in 1820 describing how a magnetic needle was affected when placed over a wire conveying an electric current (figure 14) — the first definitive proof of the reality of electromagnetism — and after he had learned in the same year of Ampère's discovery that two wires carrying currents also exert forces on each other (figure 15) — an event which led Ampère to the hypothesis that magnetism is attributable to the flow of electric charge — Faraday had brooded over the relationship between electricity and magnetism. The idea behind his own breakthrough of electromagnetic rotations in 1821 which so incensed Wollaston and Davy, came as a consequence of writing his historical survey of the evolution of electromagnetism up to April 1821. In his notebook in 1822, Faraday wrote: 'Convert magnetism into electricity.' He succeeded in

(a)

(b)

A   B

(c)

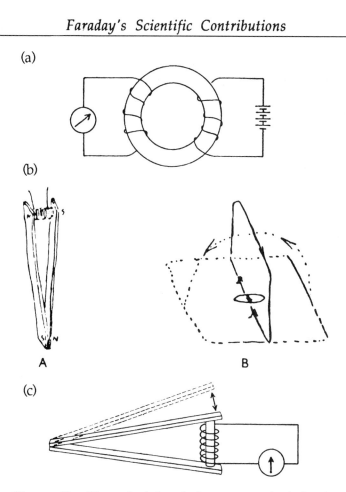

Figure 19 The principle of electromagnetic induction. Faraday discovered that when a coil on one side of a soft-iron ring was either connected to or disconnected from a battery, an electric current passed through the coil on the opposite side of the ring (part (a)). This is the principle upon which the electrical transformer works. (b) Faraday also showed that the same effects could be obtained by making or breaking the magnetic contact between two bar magnets and a coil wound on an iron cone (labelled A in the sketch from his Diary). Faraday also showed (26 December) that a single loop of wire rotated in the earth's magnetic field produces a current — see his sketch B. (c) Gives an expanded view of Faraday's sketch A.

41

August 1831, after an intense period of activity on other things. He had presented, from January to June that year, five Evening Discourses (on optical deception, light and phosphorescence, oxalamide, the production of sound during the conduction of heat) and a course of Thursday afternoon lectures in April and May.

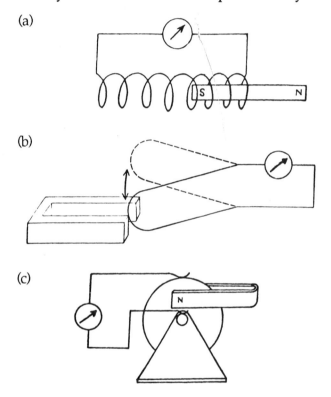

Figure 20 Faraday subsequently showed that an electric current was also produced by thrusting a magnet into a coil of wire or by withdrawing it (a), or by moving a loop of wire up and down in a magnetic field (b). He later showed that, by rotating a copper disc between the poles of a powerful magnet a steady current was induced across the disc (c).

In the space of ten days, spread over a period of ten weeks from August to November 1831, Faraday had enunciated his **laws of electromagnetic induction** that an electric current is set up in a closed circuit by a changing magnetic field (see figures 19 and 20). Thereby he laid the foundations of the modern electrical industry.

Quite apart from its other major consequences, this discovery heralded a new era in the relationship between pure research and industrial application. Industries were established in the eighteenth and nineteenth centuries built on the inventions that came from the workshop rather than from the scientific laboratory. Here was Faraday demonstrating that from the research laboratory, devoted solely to the pursuit of knowledge for its own sake, industries of incalculable importance to mankind could emerge. It is no accident that, quite soon afterwards, research laboratories soon became, and have remained ever since, the nerve centres of the large electrical companies (such as Philips, Siemens, GEC, NEC, and General Electric Schenectady).

Soon after his discovery of electromagnetic induction, Faraday began to hypothesize upon the mechanism of the phenomenon. He abhorred the idea of action at a distance implied in the operation of Coulomb's (and Cavendish's)[7] inverse square law between electrical charges and magnetic poles. He also disliked the notion that there was no role for the space in the 'vacuum' separating two objects to play in governing the forces between them, other than to provide the distance. (He had similar worries about the mode of operation of gravitational forces.)

Against this background Faraday later proposed the **concept of a field** in the spaces surrounding a magnet

(a)

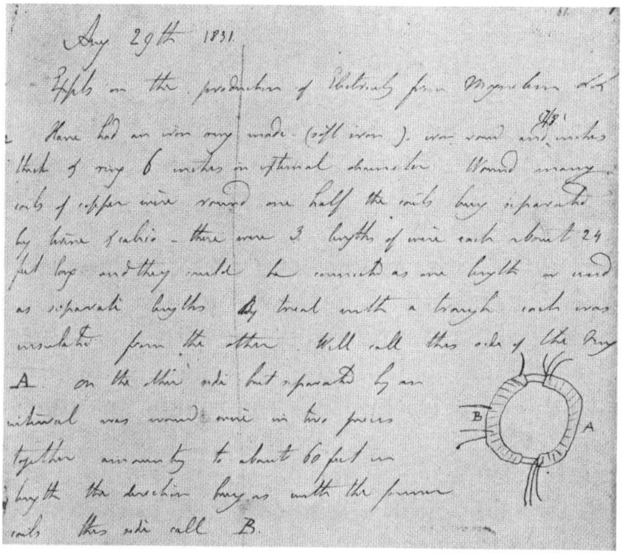

Aug 29th 1831

Expts on the production of Electricity from Magnetism &c &c

Have had an iron ring made (soft iron), iron round and 7/8 inches
thick & ring 6 inches in external diameter. Wound many
coils of copper wire round one half the coils being separated
by twine & calico – there were 3 lengths of wire each about 24
feet long and they could be connected as one length or used
as separate lengths. By trial with a trough each was
insulated from the other. Will call this side of the ring
A. On the other side but separated by an
interval was wound wire in two pieces
together amounting to about 60 feet in
length the direction being as with the former
coils this side call B.

(b)

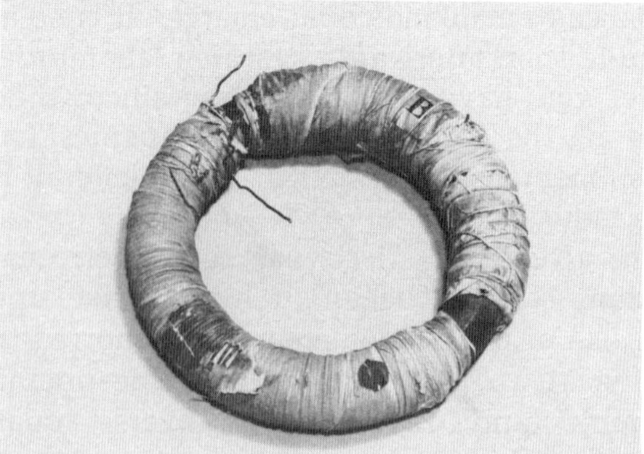

Figure 21 (a) Faraday's Diary entry for 29 August 1831. (b) The actual soft-iron ring used by Faraday.

and a wire carrying an electric current (see figure 23). With the aid of iron powder sprinkled on a thin sheet of paper under which a bar magnet was placed, Faraday demonstrated the existence of his conceived **lines of force** (figure 24). He refined his ideas on lines of force in several subsequent papers and Friday Evening Discourses over the next twenty years.

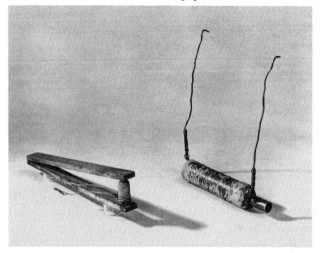

Figure 22 Photographs of original apparatus used by Faraday for the experiments schematized in figures 19 and 20.

Events were to show that this was probably Faraday's greatest contribution to physics and certainly his most important theoretical one. His lines of force ushered a new era into physics and cosmology: an era built on the concept of the field, which pervades the space around a magnet and around an electric current, and, in the words of Maxwell (much later) 'weaves a web through the sky.'

The Laws of Electrolysis

The period between January 1832 and December 1834 was exceptionally busy for Faraday. In this short time he published some twenty original papers, delivered seventeen Friday Evening Discourses (see Appendix I),

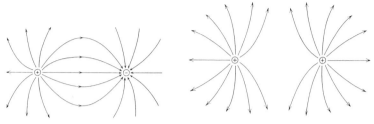

Electric field lines for a pair of unlike charges.   Electric field lines for a pair of positive charges.

Figure 23 Faraday believed that lines of force, schematized here, occupied the space separating magnetic or electric charges.

gave the (six) Christmas Lectures (on Chemistry), December 1832-January 1833, forty-eight day-time lectures on a wide range of chemical and physical topics all at the Royal Institution, delivered seventy-five lectures at the Military Academy in Woolwich, and entered into extensive correspondence with many eminent personages, including: Hachette and Ampère in Paris; Barrow,[8] the Second Secretary of the Admiralty, Charles Babbage (who devised an early forerunner of the modern computer); the Editor of the Literary Gazette (expressing his annoyance about attribution of priority as between Nobili and himself for work on electricity and magnetism, see page 103); Fazzini, the Professor of Mathematics and Metaphysics at the University of Naples, and Melloni, the famous physicist in the same city; J D Forbes in Edinburgh; Moll in Utrecht; Berzelius[9] one of the world's premier chemists in Stockholm; Plateau in Brussel;, Mary Somerville, the mathematician and popularizer of science; J F W Herschel; William Whewell, the polymath who later became Master of Trinity College, Cambridge; and a 7500 word memorandum to Gay-Lussac. And, on 11 July 1834, he appeared before the

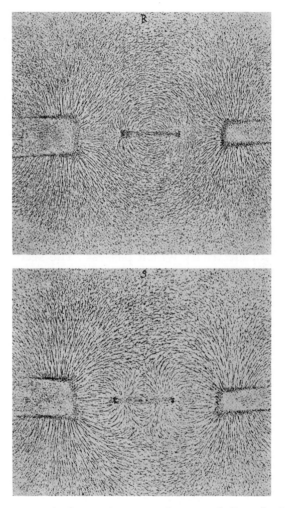

Figure 24 A favourite experiment of Faraday's to illustrate the reality of lines of force was to sprinkle iron powder on a sheet of paper beneath which there was placed a magnet. On gently tapping the paper the particles of iron 'revealed' the lines of force.

Select Committee of the Houses of Parliament to give evidence on the state of metropolis sewers. But, impressive as this catalogue of activity is, the most significant work that he accomplished in this period was his discovery of what have since become known as **Faraday's Laws** of Electrolysis.

These laws rank among the most accurate generalizations in science. All undergraduate texts in general chemistry and most in physics still give them high prominence, for reasons which will soon be apparent. They describe in quantitative terms the relationship between the extent of chemical decomposition of a conducting substance and the amount of electricity that passes through it. The first Law, in Faraday's words, states: 'Chemical action or decomposing power is exactly proportional to the quantity of electricity which passes.' The second Law, again in his own words, asserts: 'Electrochemical equivalents coincide and are the same with ordinary chemical equivalents.' In other words, the electrochemical equivalent of an element is proportional to its ordinary chemical equivalent. Today we call the amount of electricity necessary to liberate one equivalent (i.e. 1.008 grams of hydrogen and 35.457 of chlorine) from a solution of hydrogen chloride, or, in short, 1g equivalent of any element from conducting solutions or conducting molten salts of its compounds, one 'faraday' (which is 96,493 coulombs of electricity).

These Laws brought order where there was hitherto confusion. They also pinpointed the relevant factors and, conversely, the irrelevant ones. It did not matter what concentration of solution the current passed through. The nature or the dimensions of the electrodes used was also of no consequence. The key factors were simply the quantity of electricity and the chemical equivalents. These conclusions aroused deep suspicions. As Berzelius said at the time (and as Frankland was to recall in a Friday Evening Discourse much later),[10] '... compounds, whose elements were united by the most dissimilar degrees of affinity, required equal quantities of electric force for their decomposition.' In other words, nature requires the same amount of electricity to separate the binary compounds AB and CD into their elements A and B, and C and D, even though the energy released when A and B combine to form AB may be vastly different from that released when C and D combine.

The significance of this fact is that, fundamentally, chemical forces and electrical forces are intimately and quantitatively related. For example, the complete electrolysis of a solution containing one gram formula weight of barium chloride ($BaCl_2$) or of aluminium chloride ($AlCl_3$) requires double and treble the amount of electricity required to electrolyse completely a gram formula weight of hydrogen chloride (HCl). Hence atoms of hydrogen, barium and aluminium, which react with one, two and three atoms of chlorine respectively, react also with one, two and three units of electricity. As the Irish astronomer Johnstone Stoney (in 1874) and the German physiologist-turned-physicist, Hermann von Helmholtz (during the course of his Faraday lecture at the Royal Institution in 1881) concluded: electricity must have a unitary structure, the maximum value of the unit being that which is sufficient to react with one univalent atom. The full significance of Faraday's Laws of Electrolysis was not appreciated until J J Thomson, working at the Cavendish Laboratory in the 1890s, discovered the electron. All **ions**, the **species spoken of first by**

Faraday, bear a charge, and this charge occurs in multiples of the electronic charge. That matter and electricity are intimately related was not fully appreciated until the early decades of the 20th century; but the origin of the idea lies in this phase of Faraday's work.

It was Faraday, guided by Whewell, with whom he had extensive correspondence at this time, who **introduced all the major terms** nowadays used **in the language of electrochemistry**: electrolyte, electrolyse, electrolysis, electrode, anode, cathode, ion, anion and cation.

The practical consequences of Faraday's work on electrolysis were major. They led, largely in the hands of Jacobi in St Petersburg in the late 1830s, to the technique and industry of **electroplating** which had also taken root in England, France and other countries. A partnership between the cousins George and Henry Elkington and Josiah Mason at Birmingham in 1842 established that city as the world's foremost centre for **electrogilding** and **silvering**. This commercial development signalled the rather sudden demise of Old Sheffield Plate (now prized for its antique value), which was produced by heating and annealing sheet silver on to a copper-rich substratum. It could be argued that Faraday's scientific endeavours were the most important factors in shifting world dominance in the silver-plating industry from Sheffield to Birmingham. Not only was Faraday's work on electrolysis of paramount importance in this regard but his discovery of electromagnetic induction two years earlier had also led to an abundant new source of direct current electricity. As early as 1832, a young French instrument maker Hippolyte Pixii exhibited his hand-driven dynamo to the Académie des Sciences; and in

1834 a London instrument maker E Clarke developed an improved electric generator. But it was John Woolrich of Birmingham who was the first to effect large-scale electrodeposition of metals using magneto-electric machines, a fact which, with understandable civic pride, the city of Birmingham still boasts. Against one of the walls in the Chapel of Aston Hall, Birmingham stands what is claimed to be the first commercial electroplating machine. The accompanying inscription reads:[11]

> *This machine, founded upon Faraday's great discovery of Induction, was invented by the late John Stephen Woolrich of Birmingham. It was constructed by Messrs. Prime & Son in 1844, and was worked by them for many years, until superseded by machines of improved construction and greater power. It is the FIRST magnetic machine that ever deposited silver, gold or copper, and it is the forerunner of all the magnificent dynamo machines that have since been invented. Professor Faraday, on the occasion of the meeting of the British Association in Birmingham, paid a visit, together with some of his scientific friends, to Messrs. Prime & Son's Works, purposely to see the application of this great discovery in practical operation, and expressed his intense delight at witnessing his discovery so early and extensively applied and so successfully carried into practical use. To Birmingham belongs the honour not only of introducing electro-plate, the use of which has extended to every civilised nation, but also the honour of first adopting Faraday's great discovery of obtaining electricity from magnetism,— a discovery that has influenced science and art to an enormous extent.*

The practical consequences of Faraday's work on electrolysis also led to the **electroforming** industry (formerly called Galvanoplasty) in which metal is electrodeposited on a mould and then the mould is removed. This became the key process in the production of early phonograph records, in the production of medals and in electrotyping. Faraday's work also led to the ultrasensitive method of chemical analysis still widely exploited known as **electroanalysis**, where the weight of the electrodeposited material is recorded; and to **coulometry**, the measurement of the amount of electric charge transferred by a current.

Faraday had to devise and construct his own coulometers and other electrical instruments and electrochemical set ups. His so-called **volta-meter** (nowadays termed a gas-**coulometer**) is a case in point (figure 25). It enables the volume of hydrogen and oxygen produced on electrolysis of acidulated water to be precisely measured, and this, in turn, via his Laws, yields the amount of charge transferred. There was another ingenious instrument, utilizing an iodine-based meter — that had paper soaked in a starch/iodine solution and platinum electrodes — which he used, in 1833, to **establish the identity of electricity derived from different sources.**

Until that time it was questioned whether voltaic electricity was the same as electrostatic electricity — Faraday often called the latter Franklinic electricity, in deference to the pioneering work of Benjamin Franklin (1706-1790) who had, *inter alia*, shown that electricity is never created or destroyed, but only transferred. Having discovered a new method of generating electricity by moving a magnet, Faraday pondered over the further question as to whether this too, as well as animal electricity (of the kind possessed by the electric eel) and thermo-electricity, were the same. In a brilliant series of experiments he established that a standard battery of platinum and zinc connected for 3.2 seconds (eight beats of his watch) gave the same deposition (of iodine) as thirty turns of his large electrostatic generator. After this beautifully economical series of experiments, Faraday concluded that 'Electricity, whatever may be its source, is identical in its nature.' What he showed, in effect, was that the electricity of thunderstorms, the 'galvanism' of the frog's leg, the static charges stored in Leyden jars, the currents generated by a voltaic pile (which is similar to the battery of a motor car) as well as that produced by a moving magnet in a nearby wire, were all synonymous.

Figure 25 This 'volta-meter' is a cell constructed by Faraday which enables the volume of both hydrogen and oxygen produced by electrolysis to be accurately measured. The volume generated is directly proportional to the amount of electricity passed through the cell.

53

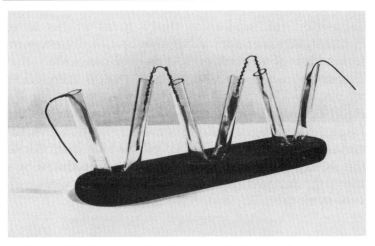

Figure 26 An example of one of the items of apparatus constructed by Faraday for his work on the electrical properties of solutions, liquids and molten solids.

In 1834, Faraday became the **first Fullerian Professor of Chemistry** at the Royal Institution thanks to the munificence of one of the sedate habitués of the place, a tall, jovial gentleman (and Member of Parliament), who, in the words of a recent commentator[12] 'lounged to Faraday's lectures in his old fashioned blue coat and brass buttons, grey smalls and white stockings.' John Fuller's generosity gave enough money to establish the chair for Faraday and another (visiting professorship) in Physiology and Comparative Anatomy. It is said that Mr Fuller,

> ... *the feebleness of whose constitution denied him at all other times and places the rest necessary for health could always find repose and even quiet slumber amid the murmuring lectures of the Royal Institution and that in gratitude for the peaceful hours thus snatched from an otherwise restless life he bequeathed to the Royal Institution the magnificent sum of £10,000!*

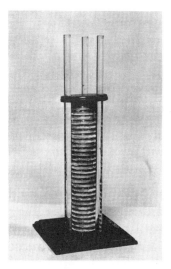

Figure 27 The Volta pile presented to Faraday by its inventor, Alessandro Volta.

In 1835, Faraday worked on an improved form of the voltaic battery, took a keen interest in Melloni's work in Italy on discoveries in radiant heat (and presented a Discourse on the subject), gave evidence to a Parliamentary committee on the prevention of dry rot, and got involved in a somewhat unpleasant set of circumstances surrounding the award of a pension.

The pension, of £300 per annum, was first proposed under the Tory government of Sir Robert Peel, but before the formalities were completed the Whigs, under Lord Melbourne, came into office. In a difficult interview with Faraday, Melbourne was reported to have referred to civil pensions as a piece of gross humbug. The facts are not known for certain but Faraday was sufficiently angered to write to Melbourne as follows:

My Lord,

The conversation with which your lordship honoured me this afternoon, including as it did, your lordship's opinion of the general character of the pensions given of late to scientific persons, induce me respectfully to decline the favour which I believe your lordship intended for me; for I feel that I could not, with satisfaction to myself, accept at your lordship's hands that which, though it has the form of approbation, is of the character which your lordship so pithily applied to it.

Faraday's friends were outraged by the news and the matter was taken up with King William IV. Melbourne wrote a letter of apology to Faraday, the pension was granted and gracefully accepted.

Faraday's relatively brief but very important contributions to the study of **electric discharges in gases** were also carried out in 1835. This is now recognized as being among the earliest work on the **chemistry and physics of plasmas**. Amongst other things it paved the way to a fuller understanding of light emission when such discharges are stuck in evacuated tubes containing noble gases. Today we know that the generation of light is a secondary phenomenon: when an electric current flows through a gas, the electrons that collide with gas atoms give up some of their energy which is then re-emitted as light.

Faraday went on to examine the **electric spark** and various forms of **discharge between conductors** of different shapes. He observed that the nature of the electric discharge between conductors inside an evacuated enclosure changed as the pressure of the residual gas decreased, noting in particular the 'dark' discharge near the cathode which is now termed the **'Faraday dark space'**.

JACK FULLER.

Figure 28 Sketch of John Fuller, that hangs in the Royal Institution. In 1833 this eccentric parliamentarian and philanthropist intimated his intention of creating a professorship of chemistry to which Faraday should be appointed, and another in physiology to which Dr Mark Roget should be appointed. (Roget is renowned for his Thesaurus.)

In 1836, Faraday pioneered work on **electrostatics** and performed his renowned cage experiment. To appreciate the background to this experiment, it is relevant to recall that Faraday, from his studies of solid conductors of all shapes and sizes, had concluded there was no such thing as an absolute charge; whenever a body is charged, an equal charge of opposite sign is induced on neighbouring bodies. Moreover, Faraday deduced (as had Cavendish long before) but left unpublished, that the charges resided on the surface of a conductor. In a dramatic experiment, performed in the lecture theatre of the Royal Institution, he sat within a hollow cube (or 'cage') of twelve-foot edge. The cube, a light wooden frame covered with conducting material, was insulated from the floor. When the cube was charged to such a degree that sparks flew from its surface, Faraday, equipped with the most sensitive instruments, could not detect any electrical action inside it. All the action was between the surface and neighbouring bodies. This gave rise to the term **Faraday cage**, which is still used to denote similar structures.

This striking demonstration prompted him to enquire what would happen if, in the space between the charged surface and neighbouring bodies, he interposed a non-conductor (that is, an insulator). His intuition led him to the view that the insulator would be set into a state of strain. This, in turn, led to his seminal studies of **dielectrics** and **specific inductive capacities**, which began a little later.

In 1836 he was appointed **Scientific Advisor to Trinity House**, a post which he much treasured and held almost up to the time of his death. It involved experimental work, both in the laboratory and extramurally, on such questions as the optimum

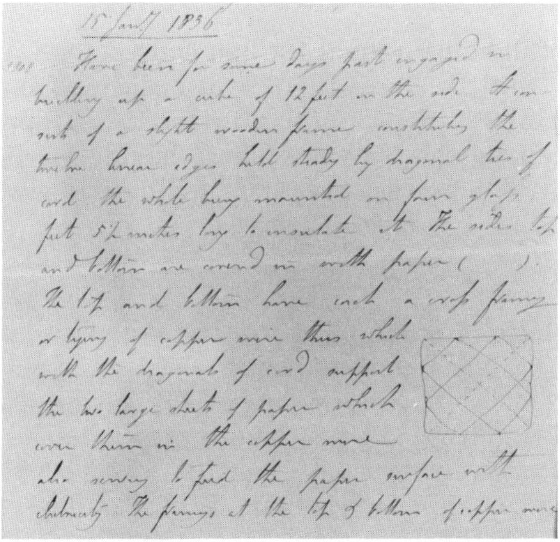

15 Jany 1836

2808 Have been for some days past engaged in
building up a cube of 12 feet in the side. It consists of a slight wooden frame constituting the
twelve linear edges held steady by diagonal ties of
cord the whole being mounted on four glass
feet 5½ inches long to insulate it. The sides top
and bottom are covered in with paper (   ).
The top and bottom have each a cross framing
or tying of copper wire thus which
with the diagonals of cord support
the two large sheets of paper which
cover them in the copper wire
also serving to feed the paper surface with
electricity The framings at the top & bottom of copper wire

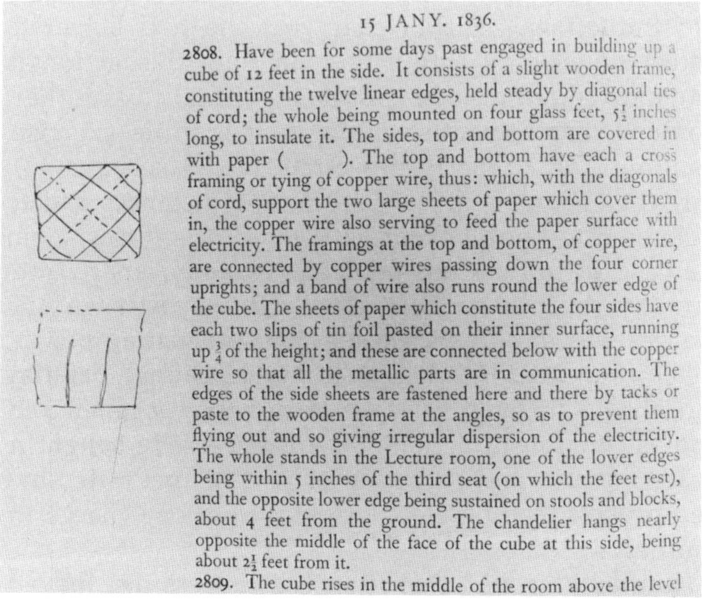

15 JANY. 1836.

2808. Have been for some days past engaged in building up a cube of 12 feet in the side. It consists of a slight wooden frame, constituting the twelve linear edges, held steady by diagonal ties of cord; the whole being mounted on four glass feet, $5\frac{1}{2}$ inches long, to insulate it. The sides, top and bottom are covered in with paper (   ). The top and bottom have each a cross framing or tying of copper wire, thus: which, with the diagonals of cord, support the two large sheets of paper which cover them in, the copper wire also serving to feed the paper surface with electricity. The framings at the top and bottom, of copper wire, are connected by copper wires passing down the four corner uprights; and a band of wire also runs round the lower edge of the cube. The sheets of paper which constitute the four sides have each two slips of tin foil pasted on their inner surface, running up $\frac{3}{4}$ of the height; and these are connected below with the copper wire so that all the metallic parts are in communication. The edges of the side sheets are fastened here and there by tacks or paste to the wooden frame at the angles, so as to prevent them flying out and so giving irregular dispersion of the electricity. The whole stands in the Lecture room, one of the lower edges being within 5 inches of the third seat (on which the feet rest), and the opposite lower edge being sustained on stools and blocks, about 4 feet from the ground. The chandelier hangs nearly opposite the middle of the face of the cube at this side, being about $2\frac{1}{2}$ feet from it.

2809. The cube rises in the middle of the room above the level

Figure 29 Excerpts from Faraday's Diary for 15 January 1836 when he constructed a cube of 12 feet for his famous 'cage' experiment.

design of lighting systems for lighthouses, fog-warning systems and the like. In 1836 also, he was made a **Member of the Senate** of the **University of London**, nominated by the Crown. Never having studied at university, this was his first involvement with a university on a semi-regular basis.

The Dielectric Constant: Experiments with Condensers: The Polarization of Molecules

On 21 December 1837, just before presenting his fifth series of (six) Christmas Lectures (on chemistry) from 28 December 1837 to 9 January 1838, Faraday read a paper of monumental importance to the Royal Society 'On Induction'. Earlier that year, he had began to reflect on electric action and on how this action is transported from one point in space to another. Contemporary scientists thought it a futile exercise. Coulomb's law was firmly established (see page 43); only the charges and their separation distances had to be known to permit the calculation of the forces, as in the case of gravitating planets. Nobody, apart from Faraday, suspected that the medium between the charges was important, except for its insulating power, which prevented the charges from dissipating. Faraday had always disliked such arguments (see page 43) and also the notion of action at a distance. He sought a clearer picture, and endeavoured to construct a step-by-step process leading electric action from one charge to another.

Faraday reasoned that if such a process did indeed exist, it would clearly be peculiar to the substance transporting the action, and so he was naturally led to experiments on the possible **influence of insulators on**

the distribution of the electric field. He made spherical condensers, as nearly identical as possible, each consisting of two concentric spheres, the inner being three centimetres less than the outer (figure 30). He first verified that if he charged only the inner sphere of one of the condensers, and afterwards brought the inner sphere of the second condenser into electrical contact with the first, the charge was shared equally between them. But then he introduced into the air gap of one of the condensers an insulator (e.g. sulphur), and by repeating the experiments with the new arrangement he made his discovery. The charge was no longer divided into two equal parts, but the condenser with the insulating material always held the greater share. Faraday measured his charges with Coulomb's torsion balance This, therefore, was a quantitative experiment and he could at once characterize the influence of the insulator by a specific constant which he called the **'specific inductive capacity'** and which nowadays is always named the **dielectric constant** or **permittivity**.[13] The unit of capacitance is now called the 'farad'.

Faraday then went on to picture how the inductive effect of an insulator can be imagined. As his substances are insulators no handing over of the charge from one molecule to another can exist. At that time Faraday was not committed to a belief in charge as we (or Cavendish and Coloumb) understand it. He also had severe reservations about molecules being made up of atoms. In what follows below a modern interpretation is given of the picture that Faraday had constructed to account for inductive effects in insulators.

[Throughout his life, Faraday seemed not to have been swayed by Daltonian arguments concerning the

Figure 30 The spherical 'condensers' used by Faraday to elucidate how an insulator stores electrical charge.

existence of atoms. This is perhaps a little surprising. John Dalton (1766-1844) gave the first public account of his atomic theory in lectures at the Royal Institution in 1803-4. In due course it was widely recognized: the French Academy of Sciences elected him a corresponding member in 1816 and later, following Davy's death, he became one of its eight foreign associates. Dalton was also elected a Fellow of the Royal Society and was one of the first to receive its Royal Medal. By the time of his death, his atomic theory had gained great popularity with many chemists but less so with experimental philosophers of a more physical persuasion. But Faraday was unconvinced, and nowhere in his writings did he find it necessary to predicate the existence of atoms. A few other eminent scientists, notably the great German chemist Wilhelm Ostwald (1853-1932) who won the Nobel Prize in chemistry in 1909, had similar feelings about atoms even up to the 1910s.]

Although no handing over of charge from one molecule to another could occur within the insulators studied by Faraday, in every molecule itself there may still be a possibility of the charges moving. If this is so, then under the influence of the electric field every **molecule will be polarized**, assuming a positive charge on one end and a negative on the other end, and all molecules pointing with their positive charges towards the negatively charged plate of the condenser. What Faraday had done was to explain how the insulator reacts on the plates of the condenser so as to draw more charge onto their surface.

Hence, from the macroscopic dielectric constant we are led to consider the molecule and its properties. Will not the mobility of the charges in different molecules be very different, and will not a connection exist between this mobility and their chemical structure? Faraday puts the question, and, as always, tries to get an answer by performing new experiments. He successively fills the space between the two spheres of one of his condensers with many gases of the most different chemical properties, and tries again and again to see if he can detect differences in the charge carried over onto the inner sphere of his second condenser filled with air. But here he fails; his method is not sensitive enough. He puts a second question: will not the temperature affect the polarizability? He tries hot and cold air, but fails again. He puts a third question: will not the polarizability of a molecule be different in different directions? He takes crystals, where molecules will be fixed in their orientation, and tries to find differences of the specific inductive capacity in different directions. He reports that in one experiment he finds such a difference, but in others cannot detect the effect.

As Peter Debye (1884-1966, the Dutch-American scientist who won the Nobel Prize in 1953 for his work on polar molecules) said in 1931 of this series of experiments '... although without positive results, Faraday's prophetic gifts are beautifully illustrated...' We know now that molecules are, for the most part, permanently positive on one side and negative on the other, and can therefore create an inductive effect by orientation alone, like small magnetic needles. The important point is, however, that Faraday felt very distinctly that his specific inductive capacity had to be intimately connected with the structure of the molecule. It is in this presentiment that he has been gloriously justified by all the work on dielectric constants up to the present time.

Exhaustion!

Through the decade of the 1830s, the strain of so much unremitting and exciting work took toll on Faraday's health. Headaches increased and lapses of memory, of which he had long suffered, worried him even more than usual. He was living 'over the shop' and there was always the temptation to work long and late. A holiday for several months in Switzerland in 1835 had revivified him somewhat, but in 1840 he suffered a complete breakdown. Mind and body were exhausted and for a whole year he rested. Many commentators claim that Faraday rested completely for the whole of 1840. But 'complete rest' for Faraday amounted to a normal workload by most people's standards. In 1840 he authored four papers in *Philosophical Magazine*, one in *Philosophical Transactions* and one in the *Literary Gazette*, covering topics such as 'voltaic precipitations', 'the source of power in the voltaic pile', 'magneto-electric induction' and 'the electricity of a jet of steam issuing from a boiler'. He also gave a course of seven lectures on chemistry that summer at the Royal Institution. By Faradaic standards, however, it was a fallow period.

During 1841 he spent eight months resting. With his wife and brother he went to Switzerland, where he took exceptionally long walks (thirty to forty miles a day). He delivered another series of Christmas Lectures (on the rudiments of chemistry) between 28 December 1841 and 8 January 1842; and in June 1841, wrote a delightful account of 'the magnificent display of lightning which we had on the 27th (of May 1841), and its peculiar appearances to crowds of observers at London' (see Chapter 5).

Magneto-optics (The Faraday Effect) and Diamagnetism

In September 1845, using a special lead borate glass that he himself had prepared almost twenty years earlier, Faraday discovered the so-called **Faraday effect**, that is the **rotation of the plane of polarization of light by a magnetic field**. This was the first ever demonstrated link between light and magnetism. It marked the **birth of magneto-optics** which, through the agency of Faraday's notion of lines of force, was to be triumphantly extended by Clerk Maxwell ten years later.

Faraday was imbued with belief in the underlying unity of the forces of nature; and it was this that prompted him to investigate the connection between light, magnetism and electricity. One of his experiments in 1845 was to observe whether plane-

Figure 31 An early photograph (*ca.* 1843) of Faraday, right, with John Frederick Daniell, the first Professor of Chemistry at King's College, London. (Daniell's electrochemical cell made him famous. It is still widely used.)

polarized light when passed through a transparent insulator was influenced by strong electric fields. He saw no evidence of any effect. Then he switched his attention to magnetism. A variety of transparent materials were placed across the poles of two powerful cylindrical electromagnets placed side by side. Rock salt, quartz, alum, fluorspar, all these, and many other transparent substances were tried, but all the results were negative. Then, with one of his own glass samples, the lead borate mentioned earlier, success came on 13 September 1845. The glasses he tried first gave, at best, only dubious results because they were so heavily strained and striated. It was when he remembered his optical glass samples that he found something good enough. So his tedious early researches ultimately paid off handsomely. As Faraday wrote '... there was an effect produced on the polarized ray and thus magnetic force and light were proved to have relations to each other.' After this result, from his arduous struggle with nature, Faraday wrote laconically in his diary: *'have got enough for today.'* When on 5th November he dispatched his paper, entitled 'On the magnetization of Light and the Illumination of Magnetic Lines of Force' to the Royal Society, the opening paragraph had celestial resonance:

I have long held an opinion, almost amounting to conviction, in common I believe with many other lovers of natural knowledge, that the various forms under which the forces of matter are made manifest have one common origin; or, in other words, are so directly related and mutually dependent, that they are convertible, as it were, one into another, and possess equivalents of power in their action. In modern times the proofs of their convertibility have been accumulated to a very considerable extent, and a commencement made of the determination of their equivalent forces.

Faraday proceeded to show that the effect first observed with the lead-containing glass, was also exhibited by many other materials, not just glass, if the conditions were right. The direction of the rotation of the plane of polarization depends upon the direction of the magnetic field — this is the Faraday effect. And it has repercussions which, to this day, are of great practical value; the tracking of spacecraft is a recent one.[14]

The phenomenon discovered by Faraday is fundamentally different from the natural rotation of the plane of polarization by certain crystals such as quartz (or by solutions of optically active molecules). When light passes through quartz and is then reflected back the resulting rotation is zero. But the rotation of the plane of polarization in any magnetized substance is doubled if the light is reflected back along the lines of magnetic force. Faraday's discovery was soon extended in different directions. The precocious William Thomson (Lord Kelvin, 1824-1907), who spent some time with Faraday at the Royal Institution — and who had written to Faraday in August 1845 describing his initial success in giving mathematical form to Faraday's notion of lines of force — took it to imply that a magnetic field encouraged rotational motion of electric charges in molecules, in accordance with Ampère's earlier suggestion that the magnetization of materials arose from the existence of little current whirls. And this idea was a source of inspiration to Maxwell as he developed his theory of electromagnetic radiation. This is where Faraday and Maxwell between them led to the hypothesis of light being an electromagnetic wave. Major conceptual advances followed when Heinrich Hertz (1857-1894) in Karlsruhe discovered electromagnetic (invisible) waves; the wonderful realization dawned that visible light is no different in principle from radiowaves. And when, in the fullness of time, the X-rays of Röntgen and the γ-rays accompanying radioactive decay were discovered, the essential synonymity of all these waves was understood. Maxwell's remark about Faraday's lines of force 'weaving a web through the sky' bore prophetic insight!

The Faraday effect has another important modern consequence: it serves as the basis of magneto-optical recording applications. During the information storage process small magnetic domains in a perpendicularly magnetized layer have their direction of magnetization reversed, while subsequent retrieval is facilitated by the optical readout of changes occurring in the linear polarization of light reflected from the various domains on the magnetic surface. Nowadays, ultra-thin films of exotic materials containing platinum, manganese and antimony are used as the medium in these information storage devices, just like the lead borate glass in Faraday's original equipment.

November 1845 was an extraordinarily productive month for Faraday. On 4th November, a day before he despatched his paper on the Faraday effect to the Royal Society, he suspended a piece of heavy glass between the poles of his new ultra-powerful horseshoe electromagnet (figure 32) and discovered that, when he switched on the current, the glass tended to set itself perpendicular to the magnetic field in the horizontal plane. The startling fact was that here was a 'non-magnetic' material — very different from lodestone or a lump of iron — showing sensitivity to a magnetic field. Rods of ordinary magnetic substances, such as iron, set themselves along the line joining the two poles of the magnet when the field was switched on. These preliminary results led Faraday to subject specimens of the utmost varied kinds to the action of his magnet: mineral salts, vegetables and animal tissues were relentlessly and excitedly examined and found to be either like iron, and so were to be called magnetic; or like glass or bismuth, and so were to be allocated to a new class of substance called **diamagnetic**. Faraday's account of his experiments has misled generations of students. The truth is that in a perfectly uniform field a rod, whether para- or diamagnetic, sets itself *parallel* to the field. But Faraday's report, like any with smallish pole pieces, had a field that became weaker away from the axis, and a diamagnetic substance is pushed towards the weaker field — that is why it sets transverse.

Figure 32 The great electromagnet constructed and used by Faraday.

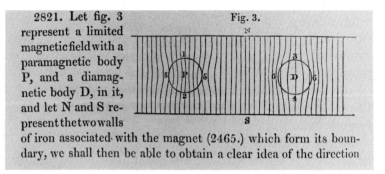

2821. Let fig. 3 represent a limited magnetic field with a paramagnetic body P, and a diamagnetic body D, in it, and let N and S represent the two walls of iron associated with the magnet (2465.) which form its boundary, we shall then be able to obtain a clear idea of the direction

Figure 33 There are more lines of force in unit volume of materials that are paramagnetic than in those that are diamagnetic. This drawing appeared in one of Faraday's publications in his series on *Experimental Researches in Electricity.*

The variety of substances examined by Faraday is very remarkable. In his account of the experiments read before the Royal Society on 18 December 1845, he remarked that he found it strange that a piece of wood, or beef, or apple should be attracted or repelled by a magnet, and went on to say that '... if a man could be suspended and placed in the magnetic field he would point equatorially' (like the glass), for all the substances of which he is formed, including the blood, had been found to be diamagnetic. To cap all this brilliant work, Faraday singled out the sluggish behaviour of a copper rod inside a magnetic fluid, and proposed an explanation in terms of **induction of electric currents** inside the copper. Such currents would nowadays be called **eddy currents,** or in France, Foucault currents.

[Léon Foucault (1819-1868), a French scientist renowned for his demonstration, with Foucault's pendulum, of the rotation of the Earth, and for the first accurate determination of the velocity of light.]

The discovery of diamagnetism was the greatest single step in founding the modern and still flourishing sub-discipline of **magnetochemistry**, the systematic study of the magnetic properties of materials. This is an indispensible tool in the armoury of the physical and biological scientist just as it is for the materials scientist and the engineer of the twentieth century.

By 1848, Faraday had identified **magnetic anisotropy** in crystals (that is, the different magnetic behaviour in a crystal according to its orientation in a magnetic field), even though he had not coined the term. It was the extension of work on magnetic anisotropy that enabled Kathleen Lonsdale and K S Krishnan, at the Davy Faraday Laboratory of the Royal Institution in the mid 1930s, to determine the orientation of molecules in crystals without the aid of X-ray crystallographic analysis.

The Night Wheatstone took Fright

It is alleged that in 1846 there occurred an event of both mythopoetic and scientific significance. This involved Faraday's friend Charles Wheatstone (see page 39). Wheatstone was due to give the Friday Evening Discourse on his electromagnetic chronoscope. He was incurably shy and panicked at the last moment leaving the Institution hurriedly. One report[15] claims that Wheatstone fled because he had heard that Joseph Crabtree, a notorious and vociferous heckler was in the audience! Faraday, who had been helping him with his preparations, stepped in and gave the Discourse himself, but realized during the course of his talk that he would finish twenty minutes before the hour was up. Reluctant to break the tradition of an exact hour's discourse, he decided to ventilate his tentative thoughts regarding the nature of light. He suggested that light could be some form of disturbance of the lines of magnetic or electric force radiating from the source. Pressed by the editor of *Philosophical Magazine* to expatiate further on this theme, he wrote his famous 'Thoughts on Ray-vibrations' which begins:

Dear Sir,

At your request I will endeavour to convey to you a notion of that which I ventured to say at the close of the last Friday evening Meeting, incidental to the account I gave of Wheatstone's electro-magnetic chronoscope; but from first to last understand that I merely threw out as a matter for speculation, the vague impressions of my mind, for I gave nothing as the result of sufficient consideration, or as the settled conviction, or even probable conclusion at which I had arrived ...

These views were greeted with much scepticism, even mild ridicule, at the time. But Maxwell in 1864 wrote 'The electromagnetic theory of light as proposed by him (Faraday) is the same in substance as that which I have begun to develop.'

Although the story concerning Wheatstone's hasty departure and Faraday's impromptu Discourse are entirely plausible, and the details in keeping with the character of, and relations between, the two protagonists, it is almost certainly apocryphal.[16] The story first gained prominence in Sylvanus P Thomson's 1901 edition of *Michael Faraday* (Cassell, London), and has been oft-repeated (see, for example, *Michael Faraday: A Biography*, by L Pearce Williams, Cambridge University Press, 1971, London). But John Tyndall (see page 153), who knows as much as anyone about the life of Faraday, made no mention of it, and neither did Bence Jones in his account of *The Life and Letters of Faraday*, (Longmans, London, 1870). What people believe to be true is often no less valid than what actually occurred. Generations of Discourse speakers have subscribed to the truth of the story, which has had the incidental effect of boosting their adrenalin — as well as that of the Director — a few minutes before their lecture begins. There is a long-standing tradition at the Royal Institution that the speaker for a Friday Evening Discourse is locked in a room for half an hour before the lecture! This adds a new meaning to the term Faraday cage.

Electricity and Gravity

Faraday tried, and failed, in 1849 to establish a link between the force of gravity and electricity. It would be quixotic to endow Faraday with a prophetic insight into this realm also. But it is justifiable to reflect that current debates about the interconvertibility of fields and matter, which *inter alia*, seek to explain the nature of subjects as seemingly disparate as the definition of the perfect vacuum and the creation of matter, all stem from the major conceptual advance that Faraday produced in formulating the notion of a field. There is considerable disagreement between historians and philosophers of science about Faraday's concept of a field. He introduced the word simply to mean the space between the poles of a magnet where the action takes place, and the development of field theory as we understand it owes more to Maxwell than to anyone else. On the other hand, it was Faraday's insistant antipathy to action at a distance that led him to promote the idea (which always remained vague) that there were processes occurring in apparently free space by which forces were transmitted. At the same time he didn't believe in aether. Maxwell and Faraday's followers (among whom Tyndall was quite specific on the matter) rejected tubes of force as the entities they came to be for Faraday, but embraced his idea of empty space as the scene, or field, of action. In other words, Faraday's private notions were an immense stimulus to his experiments, but were non-transferable to others who, however, were stimulated to take what they could use and apply it in ways Faraday could hardly have approved of, like Kelvin's and Maxwell's aether models.

The final paragraph of his paper 'On the possible relation of gravity to electricity' (*Phil. Trans.*, 1851, 1-122) has an immortal ring:

> *Here end my trials for the present. The results are negative; they do not shake my strong feeling of an existence of a relation between gravity and electricity, though they give no proof that such a relation exists.*

An attempt, very different from that pursued by Faraday's — to unify gravity and electromagnetism — came with a purely theoretical investigation by Theodor Kaluza in 1921. Kaluza showed that if Einstein's gravitational field equations are written in five dimensions instead of four (cosmologists regard the world as four dimensional: three dimensions of space, plus time) then they are precisely equivalent, when analysed from the framework of four dimensions, to the usual gravitational field equations plus the electromagnetic field equations of Maxwell (based on Faraday's work). Put differently, we may say that in some respects electomagnetism is merely a component of five-dimensional gravity. But Kaluza's theory is by no means generally accepted.

With the benefit of hindsight, we can now see how impossible it was for Faraday to achieve even a hint of unification. On the one hand, the unification of space and time in relativity theory had to wait for Einstein, and on the other the discovery of quantum physics revealed the abyss between classical ideas on gravitation and electromagnetism, and the forces operating at the fundamental particle level — weak and strong nuclear forces. In the 1930s Eddington (1882-1944, the English pioneer of stellar structures and the General Theory of Relativity) made a brave, and largely incomprehensible, effort to link the macroscopic and microscopic worlds in a single Fundamental Theory, which has been derided and forgotten. Kaluza and his supporter Klein were essentially classical, and their ideas have been resurrected by some who hope to construct a new Fundamental Theory which unites all the forces on the basis of far more knowledge than was accessible to Eddington. So there is continuity in the ideas of unification from Faraday's time to the present, but no more than that. Faraday, who was dubious about the existence of atoms and resolute against action at a distance, would have been sorely puzzled (even had he not been innocent of mathematics) by the concepts now being put forward as serious physical ideas.

The first direct experimental evidence since Faraday of a further unification in the forces of nature came from experiments using a large proton-antiproton collider at CERN in the 1980s. The giant accelerating machines at CERN permit particles of matter to collide with such force that for a fleeting moment they simulate the conditions thought to have prevailed in

Figure 34 Cartoon representation (shown by Professor Paul Davies at his Discourse on antigravity, November 1986) of Faraday's experiments to discover the link between electromagnetism and gravitation.

the primeval universe a mere billionth of a second after the initial 'big bang'.

Theoretical work in the late 1960s by Abdus Salam and Steven Weinberg had suggested an elegant link between the electromagnetic force and the so-called weak nuclear force. Their theory predicted the existence of new and very heavy subatomic particles, called W and Z. It was the discovery of these particles as predicted that convinced many members of the physics community that all the forces of nature are, at some deep level, the manifestations of a single superforce, an idea that resonates through all the writings of Faraday, as the quotation on page 67 so eloquently testifies.

In a Friday Evening Discourse given on 14 November 1986, the physicist Paul Davies, who, that night, showed the illustration reproduced in figure 34 drew attention to the grand unified theories (GUTs) that unite the strong nuclear force with the amalgamated electromagnetic-weak force of Weinberg and Salam, all these being attempts to give modern expression to Faraday's conviction of an underlying connection between gravitation and electricity.

Divided[17] Metals and Colloidal Gold

In 1857 Faraday delivered his last Bakerian Lecture on 'Experimental Relations of Gold (and other metals) to Light.' This is of major relevance in modern colloid chemistry, in the preparation and properties of comminuted (powdered) solids in general and clusters of metal atoms in particular. Although **gold sols** (that is, a system in which myriads of minute particles of gold are dispersed in an appropriate medium so as to produce a stable suspension) had been known to the alchemists of the 17th century, it was Faraday, in this paper, who was the first to make a scientific study of their preparation and properties. (This paper is also a vast repository of experiments with **aerosols** and **gels**.) He showed that the addition of salts turned the ruby-red sols blue, and then coagulated them and that these effects could be prevented by the addition of gelatin and other colloids. The ruby-coloured and (to use Faraday's adjective) amethystine sols, which had all the appearances of solutions, were shown, indirectly, by him to contain particles of gold because, unlike true solutions, a cone of light passing through them is scattered by the particles and becomes visible to an observer situated at right angles to the beam (see figure 35). (This phenomenon, later studied in detail by John Tyndall at the Royal Institution, is usually referred to as the Tyndall effect.) In retrospect,[18] we discern in Faraday's 1857 paper the embryonic stages of work that later led Zsigmondy[19], Perrin[20] and Svedberg[21] to their Nobel Prizes in the 20th century. Different gold preparations have different colours, we now realise, because of the different average sizes of the particles in the dispersion.

Faraday's 1857 paper on colloidal gold also laid some of the foundation stones of the modern physical **studies of thin films**. His experimental adventures with twelve different metals are replete with physico-chemical elegance. In one place he writes:

Hitherto it may seem that I have assumed the various preparations of (colloidal) gold, whether ruby, green, violet, or blue in colour to consist of that substance in a metallic state. I will now put together the reasons which caused me to draw that conclusion.

Figure 35 The minute particles of 'divided' gold dispersed in this preparation (by Faraday in early 1856) are responsible for the intense scattering of a laser beam.

Events, as ever, have proved him right. High-resolution electron micrography (see figure 36) of colloidal metals prepared according to the recipe given by Faraday show beyond doubt the crystalline character of the 'divided' solid. There is one other intriguing suggestion contained in Faraday's 1857 paper. Towards the end of his introduction in which he meditates on all the possibilities that relate to the interaction of gold and light, he writes: '... at one time I hoped that I had altered one coloured ray (of light) into another by means of gold.'

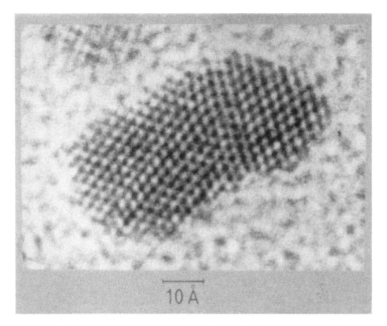

Figure 36 A high-resolution electron microscope renders visible (as black dots) the individual atoms in a dispersion of 'divided' gold. (Separation of centres of atoms is less than 0.000 000 02 cm.)

It is tempting to believe that Faraday, long before the discovery of second harmonic and sum-frequency generation, which are so much a feature of modern opto-electronics, was seeking, with his unerring intuition, to devise a means of altering the intrinsic quality of a ray by a device that would nowadays be called a frequency doubler. But we must be careful not to imply that Faraday's extraordinary intuition could have led him to phenomena completely at variance with his world-picture. He was certainly on the right lines, however, to uncover the so-called Raman effect, first identified in India in 1928. In this effect light of one wavelength gives rise to light of a well-defined longer wavelength on being scattered with the material it interacts.

[Sir C V Raman (1888-1970), Indian physicist and Nobel Laureate, who made pioneering contributions in molecular spectroscopy and acoustics, including writing treatises on many classical Indian musical instruments. Raman was for a long time a Civil Servant in Calcutta. His early scientific research was made possible by the presence near to his work and home in Calcutta, of the Indian Association for the Cultivation of Science. The laboratories and facilities there were modelled on those of the Royal Institution. He founded the Indian Academy of Science.]

In 1858, Faraday assembled his collected researches in chemistry and physics that had appeared over the preceding forty years or so in *Phil. Trans. Roy. Soc., The Journal of the Royal Institution, Philosophical Magazine* and in other publications. This collection is revealing both in what it does and does not include. His first ever paper published in 1816, — 'On the Native Caustic Lime of Tuscany' — is reproduced primarily for touching, personal reasons: Faraday records in a footnote added in 1858, 'It was the beginning of my communications to the public'; and it represents his first attempt in chemistry 'at a time when my fear was greater than my confidence, and both far greater than my knowledge.' More important is his inclusion of the whole of his (first) Bakerian Lecture to the Royal Society — delivered in three instalments in 1829 — 'On the Manufacture of Glass for Optical Purpose.' Some historians of science hold that Faraday's researches on glass constitute a relatively barren period. True, he spent a long time on them, and he even contemplated leaving the Royal Institution during the course of that work (see note 6 on page 91). But Faraday's added footnote shows him to have been proud of the crucial role his pioneering work on the preparation of glass played in his own subsequent researches into diamagnetism and magneto-optics described earlier. The optical fibres that now dominate the world of communications are often generated by spinning molten glass through small orifices in platinum-rich bushes, the very material which Faraday had so painstakingly established as the most appropriate for the production of defect-free glass.

In 1857 Faraday carried out experiments intended to **search for the role of time in magnetic effects**. He wrote to Clerk Maxwell in quite optimistic terms about the prospects of such work. No time effect was ever discovered. Other activities which occupied his mind that year included: remarks at a discussion meeting of the Institution of Civil Engineers concerning electric currents and induction in submarine telegraphs; a letter to the Dean of St Paul's and a contribution to the Report of the National Gallery site commission (of which he was a member), each concerned with the state of the marbles in the British Museum; a short article on 'The twinkling of stars' in *Philosophical Magazine*; and a letter published in the *Proc. of Roy. Med. Chir. Soc.* 'On a ready method of determining the presence, position, depth, and length of a needle broken into the foot.'

Faraday's Last Experimental Work and the Zeeman Effect

The relationship between light and magnetism was the subject of Faraday's very last experiment carried out in

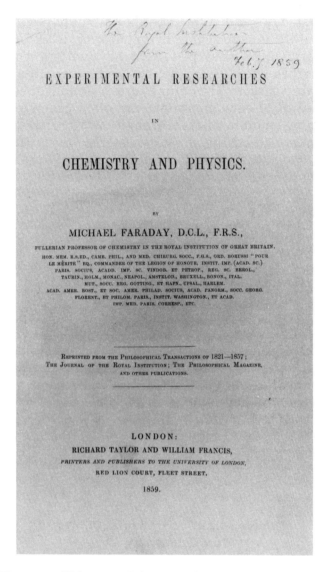

Figure 37 Title page of the copy of *Experimental Researches in Chemistry and Physics* presented to the Royal Institution by Michael Faraday.

1862. He endeavoured, but in vain, to detect any changes in the width or precise position of the lines of the spectrum of a flame (actually the bright yellow-orange colour characteristic of sodium vapour) acted on by a powerful magnet. No effect could be observed, because at that time the power of the spectroscopes, based as they were on prisms, was quite insufficient. It was not until the American worker H A Rowland (1848-1901) introduced in 1881 his concave diffraction grating that sufficient sensitivity could be achieved. And with such aids, the Dutch physicist Pieter Zeeman (1865-1943), stimulated by reading Maxwell's sketch of Faraday's great life, observed, in 1896, the phenomenon — now known as the Zeeman effect — for which Faraday was looking.

The broadening of spectral lines in a magnetic field, observed by Zeeman and elegantly interpreted by his fellow countryman Hendrik Lorentz (1853-1928) using a theorem devised by Sir Joseph Larmor, the Cambridge theoretician, played a central role in the development of quantum mechanics. The whole episode marvellously illustrates the interconnected, and often unpredictable, unexpected way in which scientific threads weave together. Lorentz, one of Europe's great theoretical physicists, said in 1918:

> *Then Zeeman could discover, as he has done, quite independently of any special theory, and without any previous consultation with myself, the magnetic resolution of spectrum lines; he could not have succeeded, if the mass of the electrons had been ten times greater with the identical electron charge.*

In 1929, Sir Oliver Lodge (1851-1940), addressing the Institute of Metals on 'States of Mind which make and miss Discoveries, with some ideas about Metals' selected the discovery of the Zeeman effect and

Faraday's original experiment to illustrate his arguments. In essence he showed that Larmor had abandoned his own attempts to seek a change in spectral feature in a magnetic field because, as a theorist, he had calculated that the effect would be vanishingly small. This was before the discovery of the electron, however. Larmor had no idea of anything smaller than an atom that was likely to radiate. Lodge remarked that:

> *Zeeman, however, undeterred by super-theory and quite independently of it, proceeded to repeat Faraday's old experiment. ... he found the effect — small indeed, but not null; the lines were slightly broadened. Directly this preliminary observation was announced, Larmor wrote to me in Liverpool asking me to repeat Zeeman's experiment, which (having a concave Rowland grating in the cupboard and a three-inch telescope) I promptly did. In about a week I had verified it, and showed the broadening of the lines at a Royal Society Soirée (20 May 1897).*

Lodge further concluded that:

> *An experimenter should seldom be deterred by a theoretical difficulty; for the data on which the theory is dependent may be erroneous. The theory mathematically may be right enough, but the data, the essential physical machinery, may be different from what had been anticipated.*

A practical application of the effect of magnetic fields upon spectral lines is the proof by its means of the existence of powerful magnetic fields in sunspots and of a general magnetic field of the sun, two major discoveries made by the American astronomer George Hale (1868-1938) at Mt Wilson in Pasadena. Hale lectured at the Royal Institution in 1909.

Faraday's Other Discoveries

In this chronological sketch of Faraday's accomplishments many of the significant observations and developments made by him have been omitted, for example his pioneering work on acoustical figures (but see pages 140-143). Some of them have since taken on a renewed importance. Only by consulting the original notebooks of Faraday does the reader capture the wealth of insight he revealed and the astonishing richness of his discoveries, even when they are, or were, regarded as minor.

If, for example, we peruse his laboratory notebook for 1833 — in the peak of one of his most fertile spells (see page 45) — we find that he discovered **fused salt electrolytes** in February of that year (figure 38), and also identified **super-ionic conductors** (silver chloride, lead chloride and some of the salts of mercury). In particular, by early November the **catalytic action of platinum** in the combination of hydrogen and oxygen had been discovered along with the **adverse effect** that **pre-adsorbed ethene** has on the catalytic activity of the metal: the first recorded example of such **retardation (or inhibition) in heterogeneous catalysis**. On 22 November, the idea had occurred to him — see entry 1096 of his Diary (figure 39) — that separation of the gas ethene (termed Oleft — an abbreviation of olefiant — by him) from carbon dioxide (which he called Carbonic acid) could, in principle, be effected by the process of **selective adsorption**. He tried spongy platinum for this purpose, and did not succeed. (Nowadays, certain porous adsorbents, such as zeolites, are used for this and related purposes.) By the 25th November, he was engaged in pioneering investigations of the influence of surface cleanliness on the **wettability of solids**. He studied the surface properties of quartz, obsidian, topaz, calcium carbonate and mica (both cleaved and uncleaved). He identified what is now called a **hydrophobic** surface, though he did not use that term.

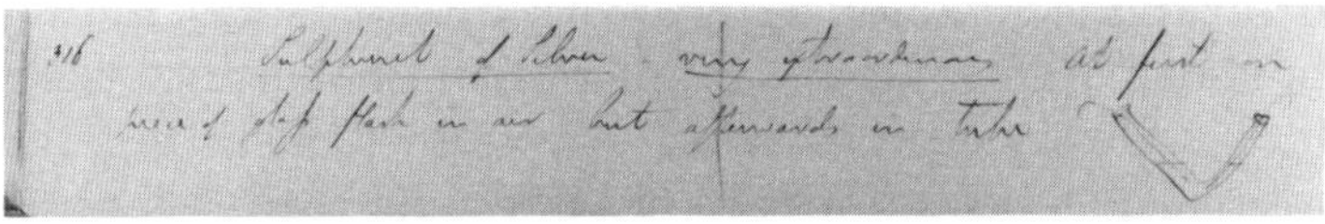

21ST FEBY. 1833.   49

316. *Sulphuret of Silver — very extraordinary.* At first on piece of glass flask in air, but afterwards in tube, fuzed into its place in fire.
317. When all was cold conducted a little (by galvanometer) and if quite cold at first conducting power did not increase. But if battery current strong from recent immersion of plates or if sulphuret warmed a little by a lamp — then as contact with poles continued the sulphuret continued to increase in conducting power as shewn by the galvanometer, and at last needle jumped into a position perpendicular to that of the coil as if on a sudden the whole electricity of battery had passed; this then continued as long as contact continued. *The heat rose as the conducting power increased* (a curious fact), no other source of heat than the current being present. Yet I do not think it became high enough to *fuze the sulphuret.* The whole passed whilst in the solid state. The hot sulphuret seems to conduct as a metal would, and could get sparks with wires at the end and a fine spark with charcoal.
318. The sulphuret when hot seems to me very much to resemble the metals in their usual state as to electrical relations, a conductor,

Figure 38 Faraday Diary entry recording the 'very extraordinary' behaviour in the electrical property of heated silver sulphide. This is the first report of thermistor action in which the electrical resistance decreases with increasing temperature.

In mid December, Faraday ruminates over the equivalence of electricity from various sources (see pages 52-53); and later that month is deeply immersed in the work that paved the way to the formulation of his Laws of Electrolysis. On the 24th December he experiments on the chemical changes wrought by the passage of the same currents of electricity through a bath of molten chloride of tin followed by another containing molten chloride of lead. His seminal experiment 'on the simultaneous and successive decomposition of chloride of lead, iodide of lead, protochloride of tin and water' was made on 26 December 1833. There was no entry for 25th December.

This dedication with his science, which became almost obsessive, is typical of his whole work (see Chapter 6).

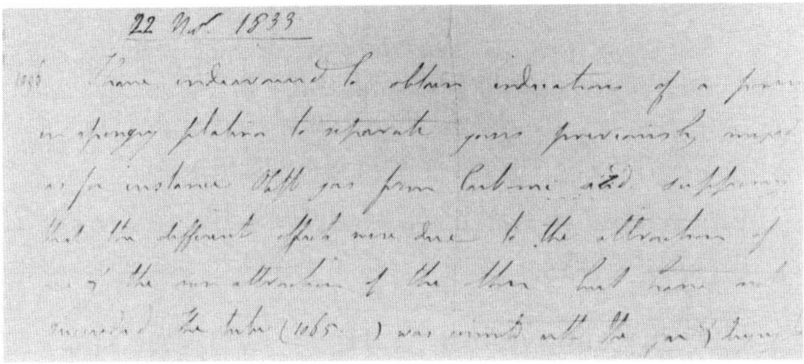

22 NOVR. 1833.

1096. I have endeavoured to obtain indications of a power in spongy platina to separate gases previously mixed, as for instance Oleft. gas from Carbonic acid, supposing that the different effects were due to the attraction of one and the non-attraction of the other, but have not succeeded. The tube (1065) was connected with the jar and drying tube there mentioned and a mixture of 1 vol. Oleft. gas with 9 vols. Carbonic acid sent through it. The gas was then expelled by distilled water as before and the Carb. acid removed by potash. The unabsorbed gas was about $\frac{1}{5}$ instead of $\frac{1}{10}$ the capacity of the tube with platina, but it did not burn very brightly, and I suspect common air is with it evolved during the solution of the C.A., etc. etc. It is very difficult in these small quantities to exclude the air from the water, etc., so as to get true results.

Figure 39 Faraday's Diary entry for 22 November 1833, recording his attempt to separate two gases by selective adsorptions (see text).

Notes

1. No other scientist has had two international units
    (the 'farad' and the 'faraday') named after him.

2. When urged by the owner of Wallsend Colliery to
    take out a patent to protect his invention which
    would have yielded a large income, Davy replied
    to him:

    > *My good friend, I never thought of such a thing; my
    > sole object was to serve the cause of humanity; and
    > if I have succeeded, I am amply rewarded in the
    > gratifying reflection of having done so ...*

    Enterprise is not always driven by the wish to
    make money.

3. His only anonymous paper.

4. Kristine Meyer — see *Supplement of Nature*, 29
    August 1931, page 339 — writes

    > *A Danish poet, a friend of Ørsted, has said
    > that he possessed what one might call a high
    > degree of 'mental innocence' and always
    > thought that 'everybody was guided by the
    > same interest for science, by the same sense of
    > justice and reason as he himself'; he thought
    > that one who knows how to deal wisely and
    > correctly will do so. One has the same
    > impression of Faraday from his life and from
    > his letters. Each man worked conscientiously
    > and unselfishly for the community in which
    > he lived.*

    See Chapters 5 and 6.

5. Nowadays called rubber. Faraday's work on rubber
    led him to describe how hydrogen sulphide
    hardened the material, a phenomenon now
    known as vulcanisation.

6. My colleague, Dr F A J L James, has pointed out
    that on Monday 25 May 1829, Lt Col C W Pasley,
    the hero of Flushing, and head of the instructional
    establishment of the Royal Engineers at Chatham,
    wrote to his colleague, Col P Drummond, Deputy
    Governor of the Royal Military Academy at
    Woolwich, concerning a replacement for a John
    MacColloch who had relinquished his position of
    chemical instructor at the Academy. Pasley said
    that the last time he had been at the Royal
    Institution he had sounded out Michael Faraday to
    see if he would be interested. Pasley remarked that
    Faraday had told him that his present salary was so
    small he did not mean to continue at the
    Institution permanently. On the previous Friday,
    Faraday had discoursed on Brunel's block making
    machine at Portsmouth, a device of more than
    academic interest to military men. It is likely,
    therefore, that this is why Drummond wrote to
    Faraday. Faraday, in replying, drove quite a hard
    bargain and his letter to Drummond, clinching the
    appointment at a figure that was double his salary
    at the Royal Institution, ended thus:

    > ... *If you could make the offer of £200 a year I would
    > undertake the lectures... I consider the offer as a
    > high honour, and beg you to feel assured of my sense
    > of it. I should have been glad to have accepted or
    > declined it independent of pecuniary motives; but
    > my time is my only estate and that which would be
    > occupied in the duty of the situation must be taken*

from what otherwise would be given to professional business.

It must not be assumed that Faraday amassed any fortune. His church preached against the accumulation of worldly riches and he spent his income as it came, chiefly on charity.

7. The eccentric Englishman Henry Cavendish (1731-1810), 'the richest of the learned, and the most learned of the rich', had anticipated Coulomb's discovery that the electrical force between charged bodies varies in inverse proportion to the square of the distance that separates them.

8. Reporting on his analysis of thirty-one samples of oatmeal suspected, and found by Faraday, to contain some 10% of calcareous matter evidently added for fraudulent purposes.

9. Faraday's letter of 5 May 1832 to Berzelius begins: 'We have been alarmed here by reports of your death but rejoiced again by finding it was false.' It ends 'That you may long live to advance chemical science and interest and delight your contemporaries is the earnest wish.' Berzelius lived another sixteen years!

10. On 2 June 1854. Edward Frankland (1825-1899), was a contemporary of Faraday's successor, John Tyndall (1820-1893). They studied together in Marburg, under R W Bunsen, of 'burner' fame. Frankland was a Professor at the Royal Institution from 1863 to 1865, when he moved to the Royal College of Science to succeed A W von Hofmann. He is best remembered for introducing the concept

of valency into chemistry, for his work with Lockyer on solar spectra that ultimately led to the discovery of helium, and for pioneering work in organometallic chemistry.

11. F Bradbury, *History of Old Sheffield Plate*, (J W Northend Ltd, Sheffield), (1983), 140.

12. G Caroe, *The Royal Institution: An Informal History*, J Murray, (1985).

13. Henry Cavendish, we now know, had discovered inductive capacity many years earlier, but he did not publish his findings.

14. As members of the Royal Institution were told in a Friday Evening Discourse on 10 November 1989 by Geoffrey Perry, former Senior Teacher at Kettering Grammar School (see *Proc. of Royal Inst. of G.B.*, **62**, (1990), 19.

15. M Goldman, *The Demon in the Aether: The Story of James Clerk Maxwell*, (Edinburgh), (1983), 90.

16. D Gooding and F A J L James (ed), in *Faraday Rediscovered*, (Macmillan), (1985), 160.

17. Both Davy and Faraday often used the term 'divided' to describe highly comminuted metals.

18. M Kerker, *Proc. of Royal Inst. of G.B.*, **61**, (1989), 229.

19. Richard Adolf Zsigmondy (1865-1929), German colloid chemist who invented the ultra-microscope. Awarded Nobel prize in 1925.

20. Jean Baptiste Perrin (1870-1942), French physicist who determined Avogadro's Number from a study of Brownian motion. Nobel prize 1926.

21. Theodor Svedberg (1884-1952), Swedish chemist who invented the ultracentrifuge. Nobel prize 1923.

Chapter 5

Faraday's Writings

Endowed with a compulsive urge to write and a memorable way of communicating his thoughts, Faraday produced a gigantic store of written contributions which, more than a century after his death, continues to reward the historian of science, kindle the hearts of the young and to strike sparks in the mind of aspiring and mature scientists alike. In addition to his four hundred and fifty original papers he also authored an early book on *Chemical Manipulation* (656pp) which ran into three editions between 1827 and 1842 and produced four volumes of his collected works; he added commentaries to them and changed the chronological sequence of the articles. One of these was his *Experimental Researches in Chemistry and Physics* published in 1859 (figure 37), the other three comprising *Experimental Researches in Electricity and Magnetism*, Vol I, II and III 1839-1855. There were two books based on his Christmas Lectures: *Various Forces of Matter*, 1860 and the immensely popular *Chemical History of a Candle*, 1861.

The manuscript records of the researches of Michael Faraday cover the years from 1820 to 1862, during which he worked in the laboratories of the Royal Institution. (These records are still in the possession of the Institution and were published[1] in 1932.) The main interest in Faraday's Diary lies not in the vast range of propositions and experimental proofs that it contains — for those have, after refinement and mature consideration by Faraday himself, been published already in various journals and elegantly embroidered in his series of *Experimental Researches* — but rather in the methods of his attack, both in thought and deed, and in the bewildering variety of problems that he tackled. His literary expression too, has a hynoptic quality: it is extraordinary that a man who did such excellent work should also have excelled in his description of it. The voluminous correspondence of Faraday — some two thousand letters of his are extant — as well as his thoughts and advice on the art of lecturing (see later) also reflect his remarkable character and shed further light on his scientific achievement.

The charm of Faraday's writing is that it gives all the details of his thought and work. He tells of his failures as well as his successes, so that the reader is tempted to believe that if he had access to a laboratory, he too might become a discoverer and be admitted to the privileged circle of those who have enlarged the bounds of human knowledge. Reading his work, one senses a unique amalgam of compelling immediacy and Chekovian timelessness, mingled with an abundance of optimism (even elation), self control and self criticism. There is no hunger after popular applause, no jealousy of the work of others, no deviation from his self-imposed practice of 'working, finishing, publishing'. His versatility, originality, intellectual energy and sheer stamina leave us in awe. There is also the sense of wonder with which, as a natural philosopher, he is imbued as he contemplates the world and the forces and mechanisms that hold it together.

The few excerpts already cited in Chapters 4 and 5 illustrate Faraday's gifts as a writer. But it is not the style alone that wins our hearts: it is the power and elegant simplicity of his arguments, which shine through as much in the magisterial sweep of the opening paragraphs of his original papers as they do in the detailed recital of factual, experimental information. Part of the magic of Faraday's writing is that it elicits admiration and conveys information in equal measure.

This brief anthology of Faraday's writings begins with his general comments on **the lecturer's art**.[2] According to Faraday:

> *His whole behaviour should evince a respect for his audience, and he should in no case forget that he is in their presence. No accident that does not interfere with their convenience should disturb his serenity or cause variation in his behaviour; he should never, if possible, turn his back to them, but should give them full reason to believe that all his powers have been exerted for their pleasure and instruction.*
>
> *The most prominent requisite to a lecturer, though perhaps not really the most important, is a good delivery; for though to all true philosophers science and nature will have charms innumerable in every dress, yet I am sorry to say that the generality of mankind cannot accompany us one short hour unless the path is strewed with flowers.*

And on the subject of **continuity** in a lecture Faraday writes:

> *I must confess that I have always found myself unable to arrange a subject as I go on, as I perceive many others do. Thus, I could not begin a letter to you on the best methods*

of renovating our correspondence and, proceeding regularly with my subject, consider each part in order and finish, by a proper conclusion, my paper and matter together.

I always find myself obliged, if my argument is of the least importance, to draw up a plan of it on paper and fill in the parts by recalling them to mind, either by association or otherwise. This done, I have a series of major and minor heads in order, and from these I work out my subject matters.

Now this method, unfortunately, though it will do very well for the mere purpose of arrangement and so forth, yet introduces a dryness and stiffness into the style of the piece composed by it; for the parts come together like bricks, one flat on the other, and though they may fit, yet they have the appearance of too much regularity. It is my wish, if possible, to become acquainted with a method by which I may write my exercise in a more natural and easy progression. I would, if possible, imitate a tree in its progression from roots to a trunk, to branches, twigs, and leaves, where every alteration is made with so much care and effect that though the manner is constantly varied, the effect is precise and determined.

On **diction and action** he gives the following advice:

In order to gain the attention of an audience (and what can be more disagreeable than the want of it?), it is necessary to pay some attention to the manner of expression. The utterance should not be rapid and hurried, and consequently unintelligible, but slow and deliberate, conveying ideas with ease from the lecturer and infusing them with clearness and readiness into the minds of the audience.

A lecturer should endeavour by all means to obtain a facility of utterance and the power of clothing his thoughts and ideas in language smooth and harmonious and at the same time simple and easy. His periods should be round, not too long or unequal; they should be complete and expressive, conveying clearly the whole of the ideas intended to be conveyed. If they are long, or obscure, or incomplete, they give rise to a degree of labour in the minds of the audience which quickly causes lassitude, indifference or even disgust.

Perhaps the reason why Faraday's fame spread so wide as a lecturer, and why, in particular, the 30 July 1851 issue of the *Texas Monument*, USA, newspaper spoke of the fluency of 'the first chemist in the world' devolved upon his strategy in **holding the interest of an audience**:

A lecturer should exert his utmost effort to gain completely the mind and attention of his audience, and irresistibly make them join in his ideas to the end of the subject. He should endeavour to raise their interest at the commencement of the lecture and by a series of imperceptible gradations, unnoticed by the company, keep it alive as long as the subject demands it. No breaks or digressions foreign to the purpose should have a place in the circumstances of the lecture, and no opportunity should be allowed to the audience in which their minds could wander from the subject.

A flame should be lighted at the commencement and kept alive with unremitting splendour to the end.

For this reason I very much disapprove of breaks in the lecture, and where they can be by any means avoided they on no account should find place. If it is unavoidably necessary to complete the arrangement of some experiment, or for other reasons, leave some experiment in a state of progression, employ as much as possible the minds of the audience during the unoccupied space — but if possible, avoid it.

Digressions and wanderings produce more or less the effects of a complete break or delay in the lecture, and should therefore never be allowed except in very peculiar circumstances; they take the audience from the main subject and you then have the labour of bringing them back again (if possible).

For the same reason (namely, that the audience should not grow tired) I disapprove of long lectures. One hour is enough for anyone, and they should not be allowed to exceed that time.[3]

Turning to Faraday's original papers, two excerpts exemplify the effective way he **introduces the reader to his subject.**[4] The first of these is the opening of his 1837 paper 'On Induction'.

Induction: an action of contiguous particles.

The science of electricity is that state in which every part of it requires experimental investigation; not merely for the discovery of new effects, but what is just now of far more importance, the development of the means by which the old effects are produced, and the consequent more accurate determination of the first principles of action of the most extraordinary and universal power in nature:— and to those philosophers who pursue the inquiry zealously yet cautiously, combining experiment with analogy, suspicious of their preconceived notions, paying more respect to a fact than a theory, not too hasty to generalize, and above all things, willing at every step to cross-examine their own opinions, both by reasoning and experiment, no branch of knowledge can afford so fine and ready a field for discovery as this. Such is most abundantly shown to be the case by the progress which electricity has made in the last thirty years: Chemistry and Magnetism have successively acknowledged its over-ruling influence; and it is probable that every effect depending upon the powers of inorganic matter, and perhaps most of those related to vegetable and animal life, will ultimately be found subordinate to it.

Amongst the actions of different kinds into which electricity has conventionally been subdivided, there is, I think, none which excels, or even equals in importance, that called Induction. It is of the most general influence in electrical phenomena, appearing to be concerned in every one of them, and has in reality the character of a first, essential, and fundamental principle. Its comprehension is so important, that I think we cannot proceed much further in the investigation of the laws of electricity without a more thorough understanding of its nature; how otherwise can we hope to comprehend the harmony and even unity of action which doubtless governs electrical excitement by friction, by chemical means, by heat, by magnetic influence, by evaporation, and even by the living being?

Twenty years later, Faraday began his article 'On the Conservation of Forces' thus:

Various circumstances induce me at the present moment to put forth a consideration regarding the conservation of force. I do not suppose that I can utter any truth respecting it that has not already presented itself to the high and piercing intellects which move within the exalted regions of science; but the course of my own investigations and views makes me think that the consideration may be of service to those persevering labourers (amongst whom I endeavour to class myself), who, occupied in the comparison of physical ideas with fundamental principles, and continually sustaining and aiding themselves by experiment and observation, delight to labour for the advance of natural knowledge, and strive to follow it into undiscovered regions.

There is no question which lies closer to the root of all physical knowledge, than that which inquires whether force can be destroyed or not. The progress of the strict science of modern times has tended more and more to produce the conviction that 'force can neither be created nor destroyed,' and to render daily more manifest the value of the knowledge of that truth in experimental research. To admit, indeed, that force may be destructible or can altogether disappear, would be to admit that matter could be uncreated; for we know matter only by its forces: and though one of these is most commonly referred to, namely gravity, to prove its presence, it is not because gravity has any pretension, or any exemption amongst the forms of force, as regards the principle of conservation; but simply that being, as far as we perceive, inconvertible in its nature and unchangeable in its manifestation, it offers an unchanging test of the matter which we recognize by it.

His early mastery in the economical handling of **factual experimental information** is demonstrated in this description of the 'white pulverulent substance' formed by the 'Combinations of Ammonia with Chlorides' (1818):

Exposed to the atmosphere, it deliquesces, but not so rapidly as chloride of lime. Thrown into water it dissolves, forming a strong alkaline solution. Heated, it gives off ammonia, and the chloride remains unchanged. Placed in chlorine it inflames spontaneously, and burns with a pale yellow flame.

His **optimism**, too, shone through early in his career as a chemist, as the final paragraph of his 1820 paper on chloro- and iodo- compounds of carbon shows:

As yet, I have not succeeded in procuring an iodide of carbon, but I intend to pursue these experiments in a bright season of the year, and expect to obtain this compound.

He never did!

From Faraday's **extensive correspondence** we capture him in a variety of moods. His righteous but controlled indignation emerges from his letter to the Editor of the *Literary Gazette*, 27 March 1832:

My Dear Sir,

Will you let me call your attention for a moment to the Article Electricity & Magnetism *at p 185 of your last Gazette? You there give an account of Nobili's experiments and speak of them as if independent of or any thing but a repetition of mine. But if you had seen Nobilis paper you would have found that my name is on every page; that the experiments in it were a consequence of his having seen a copy of my letter to Paris, which letter he translates into Italian & inserts; & that he tried and obtained the spark with the magnet, because in my letter, I said that I had obtained the spark in a particular case. Nobili, so far from wishing to imply that the experiments & discovery are his; honors me by speaking of the 'nuove correnti di Faraday'.*

I should not have noticed the matter but that in the Gazette it is said 'researches of Mr Faraday which were rapidly tending to the same discovery' whereas they are my own experiments which having gone first to Paris & then to Italy have been repeated & studied by Signori Nobili and Antinori.

Perhaps the mistake may have risen from the circumstance of the Number of the Antologia bearing date November 1831. But that date is no guide for the work was not published, or printed even, till long after that; & Nobili's paper in it is dated January 1832.

Excuse my troubling you with this letter, but I never took more pains to be quite independant of other persons that in the present investigation; and I have never been more annoyed about any paper than the present by the variety of circumstances which have arisen seeming to imply that I had been anticipated.

Gracious acknowledgment is conveyed to Mary Somerville, Oxford, 1 March 1834:

Dear Madam,

I cannot refuse myself the pleasure any longer of thanking you for your kindness in sending me a copy of

*your work. I did intend to read it through first; but I
cannot proceed so fast as I wish because of constant
occupation.*

*I cannot resist saying too what pleasure I feel in your
approbation of my late Experimental Researches. The
approval of one judge is to me more stimulating than the
applause of thousands that cannot understand the
subject.*

And also to William Whewell, Cambridge, later that
year (see page 46):

My Dear Sir,

*I ought before this to have thanked you for your great
kindness in the matter of the names respecting which I
applied to you; but I hoped to have met you last
Saturday at Kensington and therefore delayed
expressing my obligations.*

*I have taken your advice and the names used are
anode cathode anions cations and ions the last I shall
have but little occasion for. I had some hot objections
made to them here and found myself very much in the
condition of the man with his son and Ass who tried to
please every body; but when I held up the shield of your
authority it was wonderful to observe how the tone of
objection melted away.*

*I am quite delighted with the facility of expression
which the new terms give me and shall ever be your
debtor for the kind assistance you have given me.*

His self-discipline and style in declining invitations
to socialize is represented by his response to Sir John
Rennie (1794-1874), the engineer who constructed
London Bridge:

Dear Sir,

*I am very much obliged to you for your kind invitation
but am under the necessity of declining it because of a
general rule which I may not depart from without*

offending many kind friends.— I never dine out except with our Presidents — the Duke of Sussex or the Duke of Somerset whose invitations I consider as commands. Under these circumstances I hope you will accept my obligations to you though I cannot accept your favour.

That Faraday was able, notwithstanding his seemingly unending regimen of scientific work and religious commitments, to cultivate profound friendships and contentedness is apparent in a letter that he wrote to Professor Auguste de la Rive (whom he first met as a young man on his grand European tour) in Geneva shortly after reaching the age of sixty (16 October 1852). It also sheds some light on the health difficulties that he experienced at this time:

My Dear De La Rive,

From day to day and week to week I put off writing to you, just because I do not feel spirit enough; not that I am dull or low in mind, but I am as it were becoming torpid: a very natural consequence of that land of mental fogginess, which is the inevitable consequence of a gradually failing memory. I often wonder to think of the different causes (naturally) of different individuals, and how they are brought on their way to the end of this life. Some with minds that grow brighter and brighter but then physical powers fail; as in our friend Arago; of whom I have heard very lately by a nephew who saw him on the same day in bed and at the Academy: such is his indomitable spirit.— Others fail in mind first, whilst the body remains strong, others again fail in both together: and others fail partially in some faculty or portion of the mental powers, of the importance of which they were hardly conscious until it failed them. We may, in our course through life, distinguish numerous cases of these and other natures; and it is very interesting to observe the influence of the respective circumstances upon the characters of the parties and in what way these circumstances bear upon their happiness. It may seem very trite to say that content appears to me to be the great compensation for these various cases of natural change; and yet it is forced upon me, as a piece of knowledge that I have ever to call afresh to mind, both by my own spontaneous and unconsidered desires and by what I see in others. No remaining gifts though of the highest kind; no grateful remembrance of those which we have had, suffice to make us well and be content under the sense of the removal of the heart of these which we have been conscious of.

I wonder why I write all this to you: Believe me it is only because some expressions of yours at different times, make me esteem you as a thoughtful man and a true friend.— I often have to call such things to remembrance in the cause of my own self examination and I think they make me happier. Do not for a moment suppose that I am unhappy. I am occasionally dull in spirit but not unhappy; there is a hope which is an abundantly sufficient remedy for that, and as that hope does not depend on ourselves, I am bold enough to rejoice in that I may have it...

The best synthesis of Faraday's style, scientific substance and philosophical and poetic musings is to be found in the summaries of Friday Evening Discourses that he published in the *Proceedings of the Royal Institution*. The one he wrote of the Discourse 'On the Magnetic Characters and Relations of Oxygen and Nitrogen' is a particularly striking example. It also reminds us of the enormous care that Faraday took in selecting and perfecting his demonstrations, and of the element of desirable showmanship associated with some of his most telling public experiments.

Friday Evening Discourse, 24 January 1851
Sir R.I. Murchison, Vice-President, in the Chair.

Professor Faraday
On the Magnetic Characters and Relations of
Oxygen and Nitrogen.

In a Friday Evening discourse on the diamagnetic condition of flame and gases, delivered on the 14th April, 1848, Mr. Faraday called attention to the singular condition of oxygen gas in its relation to the magnet. It was then demonstrated that this gas was magnetic by its carrying a cloud of muriate of ammonia (itself diamagnetic) to the poles of the magnet, around which it seemed to gyrate in vortices. A more elaborate paper on the same subject had previously appeared in the Phil. Mag. for December, 1847.

Last year M. Becquerel, not aware of these researches, had rediscovered the high magnetic character of oxygen, made some independent investigations, and derived numerical results from them. These inquiries Mr. Faraday does not consider to interfere with, but strongly to confirm his own.

Oxygen is one of the most remarkable of known bodies: it forms one half of the aggregate of all matter. Important as are its magnetic properties, it seems incapable of receiving permanent magnetism like steel or the natural loadstone.— By a series of elementary experiments the audience were led to discriminate between these bodies, and soft iron, nickel, cobalt; which unless while under an extraneous magnetic influence, have no attractive force. Oxygen being of the latter class, it is not certain that, even while it possesses an attractive power, it is in the exact condition of the permanently magnetical body from which it derives it.

Were oxygen highly magnetic in the same extent as iron is, the immense quantity of magnetic power which would in that case be constantly undergoing variation by combustion, respiration, &c., would cause the most serious disturbances in nature. It is necessary to the conservation of the present state of things that the magnetic power in a given bulk of oxygen should be comparatively small. The audience were therefore told to expect no great demonstration of magnetism; but the extent to which that power does exist in oxygen and air, was proved by the following experiments:—

A double cone of iron (the apices of the cones meeting in a point and the cones being equal and similar,) was fabricated of such a length as to complete the magnetic circuit when placed between the poles of the large electro-magnet in possession of the Royal Institution.[5] Mr. Faraday directed attention to this hourglass-shaped piece, and showed how, by such an arrangement, extreme power is exerted at the place without any chance of change in the form of the parts. Very small soap-bubbles were blown by means of a glass tube drawn to a fine point, from a bladder filled with oxygen. It was observed that these bladders so filled were drawn forcibly inwards to the apices of the cones, but that no such effect followed when bubbles were filled with nitrogen. Another experiment, which was visible all over the room, at once demonstrated the same fact, and illustrated a differential mode of measuring the magnetic force of oxygen. A delicately balanced wire was suspended from its centre of gravity by 10 fibres of the cocoon of the silkworm; from the extremities of a small cross bar at one end of this wire were hung small glass bubbles; and the whole was so adjusted that the bubbles were on opposite sides of the apices first described, each hanging near to it but not in contact with the iron, and each equidistant from it. Therefore any difference of magnetic influence on the bubbles or their contents would be indicated by the bubble so affected being drawn inwards. In order to render such a motion widely visible, the other arm of the balance just described was converted into a long indicating lever, constructed of a straw for the sake of *lightness. To the extremity of the longer end a slip of silk was attached to catch the eye, and the lever was shielded from the currents in the room by being placed within a glass balloon two feet in diameter. By the motion of the lever it was seen, when one of the bubbles was filled completely or partially with oxygen and the other with nitrogen, that nitrogen, whether dense or rare, was totally unaffected by the magnet, and that oxygen was magnetic in direct proportion to its density in the bubble; and that the force required to set the bubble of oxygen (one atmosphere) in motion towards the magnet was one-tenth of a grain for one-third of a cubic inch of oxygen.*

Certain peculiarities in the exertion of the power which is here in action, not as a central, but as an axial force, were then referred to.

The inference from the experiment, supported by other experiments on bubbles containing air, is — that as oxygen enters into the atmosphere in a constant proportion, and as the magnetic power of oxygen varies directly with its density, definite variation must take place in the magnetic power of the atmosphere in different states.

Mr. Faraday was led to inquire whether any separation of oxygen from nitrogen in a mixture of these gases could take place, as happens when a magnet is presented to a mixture of iron filings and sand. To test this idea he applied to the conical angle (so often described as the centre of magnetic force,) a glass tube drawn to a point (as in the apparatus used for blowing the delicate soap-bubbles,) and filled with water; by slowly withdrawing the water, the air could be drawn into the tube from any desired spot and tested. This was done; and it was found that even when the magnetic action was most intense, the proportions of the magnetic oxygen and of the non-magnetic nitrogen were undisturbed.— The following experiment proved that no condensation was produced on oxygen by the magnetic power, i.e. that it is not aggregated, as happens with iron filings when under the influence of the magnet. The flat-faced poles of the magnet were separated the 60th of an inch by a copper plate with an aperture in the middle, so that when the whole was clamped together a chamber was formed. By gauges attached to this chamber it was found that no trace of condensation occurred, however great was the magnetical force brought to bear on the oxygen.

The loss of magnetical power occasioned by heat was then noticed. This was shown first in the case of iron heated to redness; then in that of nickel raised to the temperature of boiling oil; and lastly, in the case of air (i.e. of the oxygen in this air) by the following experiment:— Two conical poles a little separated were employed; above was placed a piece of phosphorus on paper, and below a helix of platinum wire heated to redness by a small Grove's battery [6] independent of that used to excite the electro-magnet. The heated air, rising upwards from the helix, speedily inflamed the phosphorus above it whilst the electro-magnet was unexcited; but when rendered active, the oxygen in the heated air becoming less magnetic, was displaced by the current of colder (and consequently more magnetical) oxygen, and the phosphorus in consequence remained unaffected by the mass which glowed beneath it, until the electro-magnet was deprived of its power; and then the natural laws of specific gravity came again into operation, the heated air rose, and the phosphorus was lighted.

In conclusion, Mr. Faraday announced his intention of applying, on a future evening, the reasoning deducible from these and other experiments, to the variation of magnetic lines on the earth's surface. His purpose then will be to compare the records of this varying force with the variations of temperature occasioned by the annual revolution of the earth, the varying pressure of the atmosphere, storms, &c. with the hope of supplying a true theory of the cause of the annual and diurnal, and many of the irregular variations of the terrestrial magnetic power.

For the papers in which these results are described more at large, see Philosophical Magazine, 1847, Vol. XXXI. p. 401; and Philosophical Transactions for 1851, p.1.

The chairman for the above Discourse, Rhoderick Impey Murchison (1792-1871), was the famous Scottish geologist who, in his early thirties, gave up fox-hunting for geological science. After some geological work in Scotland alongside Adam Sedgwick and Charles Lyell, Murchison turned his attention to the 'greywacke' rocks underlying the Old Red Sandstone. His classic work in South Wales and the Welsh Borderland led to his monumental tome, *The Silurian System*, published in 1839. (The Silures were an ancient Welsh tribe.) Murchison became Director-General of the Geological Survey in 1855. He was one of the founders of the British Association and was its President in 1846. In 1860, he and Faraday reported on processes for preserving the external stonework of the Houses of Parliament. Faraday took great delight in observing natural phenomena as his diary entries during his visits to Wales and Switzerland testify. At the time of his recuperation from physical and nervous exhaustion, he wrote the following note to the Editor of *Philos. Mag.* in June 1841:

On some supposed forms of Lightning.

The magnificent display of lightning which we had on the evening of the 27th of last month, and its peculiar appearance to crowds of observers at London, with the consequent impressions on their minds, induce me to trouble you with a brief letter on certain supposed appearances and forms of lightning, respecting which the judgement of even good observers is often in error.

When, after a serene sky, or one that is not overcast, thunder-clouds form in the distance, the observer sees the clouds and the illumination of the lightning displayed before him as a magnificent picture; and what he often takes to be forked lightning (i.e. the actual flash, and not a reflexion of it), appears to run through the clouds in the most beautiful manner. This was the case on that evening to those who, being in London, observed the storm in the west, about nine o'clock, when the clouds were at a distance of twenty miles or more; and I have very frequently observed the same effect from our southern coasts over the sea. In many of these cases, that which is thought to be the electric discharge is only the illuminated edge of a cloud, beyond and behind which the real discharge occurs. It is in its nature like the bright enlightened edge which a dark well-defined cloud often presents when between the sun and the observer; and even the moon also frequently produces similar appearances. In the case of its production by lightning and distant clouds, the line is so bright by comparison with the previous state of the clouds and sky, so sudden and brief in its existence, so perfectly defined, and of such a form, as to lead every one at the first moment to think it is the lightning itself which appears.

But the forms which this line assumes, being dependent on the forms of the clouds, vary much, and have led to many mistakes about the shape of the lightning flash. Often, when the lightning is supposed to be seen darting from one cloud to another, it is only this illuminated edge which the observer sees. On other occasions, when he was sure he saw it ascend, it was simply this line more brilliant at its upper than at its lower part. Some writers have described curved flashes of lightning, the electric fluid having parted from the clouds, gone obliquely downwards to the sea, and then turned upwards to the clouds again: this effect I have occasionally seen, and have always found it to be merely the illuminated edge of a cloud.

I have seen cases of this kind in which the flash appeared to divide in its course, one stream separating into two; and when flashes seen at a distance are supposed to exhibit this rare condition, it is very important the observer should be aware of this very probable cause of deception.

I have also frequently seen, and others with me, a flash having an apparently sensible duration, as if it were a momentary stream, rather than that sudden, brief flash which the electric spark always presents, whose duration even Wheatstone could not appreciate. This I attribute to two or three flashes occurring very suddenly in succession at the same place, or nearly so, and illuminating the same edge of a cloud.

The effect I have described can frequently be easily traced to its cause, and when thus traced best prepares the mind to appreciate the mistakes it may lead, and has led, to in the character, shape and condition of the lightning flash. It often happens at the sea-side, that, after a fine day, clouds will towards evening collect over the sea on the horizon, and lightning will flash about and amongst them, recurring at intervals as short as two or three seconds, for an hour or more together. At such times the observer may think he sees the lightning of a flash; but if he waits till the next illumination, or some future one, takes place, he will perceive that the flash appears a second time in the same place, and with the same form; or perhaps it has travelled a little distance to the left or right, and yet has the same form as before. Sometimes an apparent flash, having the same shape, has occurred three or four times in succession; and sometimes it has happened that a certain shaped flash having appeared in a certain place, other flashes have appeared in other places, then the first has reappeared in its place, and even the others again in their places. Now in all these cases it was simply the illuminated edges of clouds that were seen, and not the real flashes of lightning. These forms frequently exist in the cloud, and yet are not distinguishable till the lightning occurs. It is easy, however, to understand why they are then only developed, for that which appears in the distance to be one dull mass of cloud, distinguishable in figure only at its principal outline, often consists of many subordinate and well-shaped masses, which, when the lightning occurs amongst or beyond them, present forms and lines before unperceived.

The apparent duration, which I before spoke of, is merely a case of very rapidly recurring flashes, and may, by a careful observer, be easily connected with that which I have now proposed as the best test of the nature of the phenomena.

There are some other circumstances which will help to distinguish the effect I have thus endeavoured to describe from the true appearance of the lightning flash, as the apparent thickness, sometimes, of the supposed flash, and its degree of illumination; but I have, I think, said enough to call attention to the point; and, considering how often the philosopher is, in respect to the character of these appearances, obliged to depend upon the report of casual observers, the tendency of whose minds is generally rather to give way to their surprise than to simplify what may seem remarkable, I hope I have not said too much.

There is little doubt that Faraday's felicity of expression, exemplified by the above passages, contributed to his genius.

Notes

1. *Faraday's Diary* (1820-1862), Vol I-VII. (Ed T Martin, Foreword by Sir William H Bragg.) G Bell and Sons, 1932. Faraday did not commence numbering his paragraphs until, providentially, 29 August 1831, the date of his discovery of electromagnetic induction. From that date to his last entry, on 12 March 1862, there are some 17,000 paragraphs.

2. The Royal Institution has published several booklets on this topic, one edited by the late Geoffrey Parr being a compilation of Faraday's *Advice to a Lecturer*, the other bearing the same title, by Sir George Porter and James Friday, an anthology taken from the writings of Michael Faraday and Lawrence Bragg. I M McCabe and J M Thomas have written about and cited a selection of Sir Lawrence Bragg's lecturing skills in their article in *Education in Chemistry*, **27**, 156-159, (1990) entitled 'The Legacy of Lawrence Bragg'.

3. Friday Evening Discourses at the Royal Institution last precisely one hour.

4. See also page 67, the opening of his paper on the discovery of the Faraday effect.

5. See figure 32.

6. A more powerful source of voltaic electricity than the conventional pile (see page 147 and figure 48).

Chapter 6

Faraday the Man

An analysis of Faraday's life reveals a number of interesting paradoxes. Though a public figure, known to all the famous people and the majority of the general public of the Victorian era, he remained throughout his life a private person uninterested in social ambition and in the pursuit or attractions of wealth. Festooned with honorary degrees from numerous universities and elected a member of all the premier academies of the world (see Appendix II), and proud to acknowledge such honours after his name in the title pages of the books he wrote (see figure 37), he nevertheless shunned publicity and instructed that his gravestone should read simply, Michael Faraday. Though he frequented the occasional Soirée, held in the salon of the wealthy philanthropist, Baroness Burdett Coutts[1] with whom he had considerable rapport, or in Great Marlborough Street where he mingled with artists and musicians (including the painter J W M Turner), he was happier in the privacy of his own home, and in the company of his wife, niece and brother. Serene in the security of his religious conviction, he was untroubled by the apparent conflict between science and religious beliefs. He could excoriate the spiritualists for their naivety of the faith (see pages 125-127) while at the same time accept, as did his fellow Sandemanians, the literal truth of the Bible. Resolute in his pursuit of excellence as a lecturer and dedicated to the attainment of the highest standards in the Royal Institution at Albemarle St, he accepted with equanimity the primitive theological pronouncements of his fellow worshippers in the Paul's Alley Meeting House [2] off Aldersgate St.

His successor John Tyndall reveals much about Faraday the man:[3]

> *We have heard much of Faraday's gentleness and sweetness and tenderness. It is all true, but it is very incomplete. You cannot resolve a powerful nature into these elements, and Faraday's character would have been less admirable than it was had it not embraced forces and tendencies to which the silky adjectives 'gentle' and 'tender' would by no means apply. Underneath his sweetness and gentleness was the heat of a volcano. He was a man of excitable and fiery nature; but through high self-discipline he had converted the fire into a central glow and motive power of life, instead of permitting it to waste itself in useless passion. 'He that is slow to anger', saith the sage, 'is greater than the mighty, and he that ruleth his own spirit than he that taketh a city.' Faraday was not slow to anger, but he completely ruled his own spirit, and thus, though he took no cities, he captivated all hearts.*

Tyndall poses the question:

> *Is he not all the more admirable, through his ability to tone down and subdue that fire and that excitability, so as to render himself able to write thus as a little child? I once took the liberty of censuring the closure of a letter of his to the Dean of St. Paul's. He subscribed himself, 'humbly yours', and I objected to the adverb. 'Well, but, Tyndall', he said, 'I am humble; and still it would be a*

great mistake to think that I am not also proud.' This duality ran through his character. A democrat in his defiance of all authority which unfairly limited his freedom of thought, and still ready to stoop in reverence to all that was really worthy of reverence, in the customs of the world or the characters of men.

That he was earnest and studious is abundantly apparent from the wealth of information disclosed in his notebooks, correspondence, published papers and evidence given at public enquiries and select committees. But he was also capable of enjoying himself in a variety of rather simple ways. His rich imagination pertained to things other than the scientific. He once described himself when a young apprentice as 'a very lively imaginative person' who 'could believe in the *Arabian Nights* as easily as in the *Encyclopaedia*; but facts were important to me. I could trust a fact.'

The beauty of nature, especially the hills of Devonshire, the vales of South Wales, all the Alpine landscapes and the seascapes of Brighton or the Isle of Wight, could move him to lyrical ecstasy. And in contemplating waterfalls, the rainbow or lightning, his responses were often Wordsworthian, though never expressed in verse. Music and the theatre appealed to him; and so did the craftsmanship required in attending to his hobbies as a collector and to the custodial side of his duties as the Superintendent of the House of the Royal Institution. On quiet evenings he probably attended to attaching his nameplates (figure 40) on his furniture[4] or to sticking interesting letters and cuttings into large albums (figure 41), with lithographs and engravings for illustrations.

The Faradays were childless, but they had a blissfully happy married life. Sarah Barnard never shared the

Figure 40 Faraday took delight in attaching brass nameplates to his furniture, many pieces of which are still in use at the Royal Institution.

scientific interests that drove Faraday with such febrile intensity: she said she was happy to be 'the pillow of his mind'. The company of young people, especially his nieces, Jane and Constance Reid, who lived for a few years with the Faradays at the Royal Institution, meant a great deal to him. Occasionally the party went up the river Thames in a rowing boat to picnic. Dafydd Tomos' *Faraday in Wales*, tells of a delightful incident in the Vale of Neath when Faraday engaged himself in conversation with a ten-year-old girl. He was struck by the fact that he could understand her, and appreciate all her winning charm, even though she spoke to him in a foreign tongue. His niece has recalled how, when she visited him in the laboratory he would drop a piece of potassium into a trough of water for her to watch it dart across the surface. She has also told of how, at children's parties, Faraday rode a velocipede in the corridor round the back of the lecture theatre! He would occasionally take some physical exercise[5] by cycling on his velocipede as far as Hampstead. The

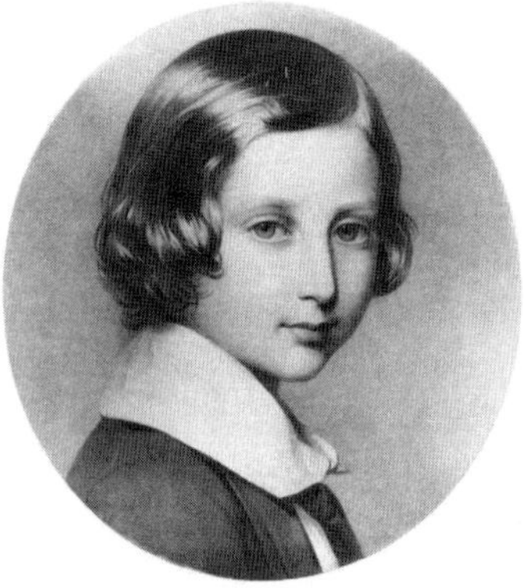

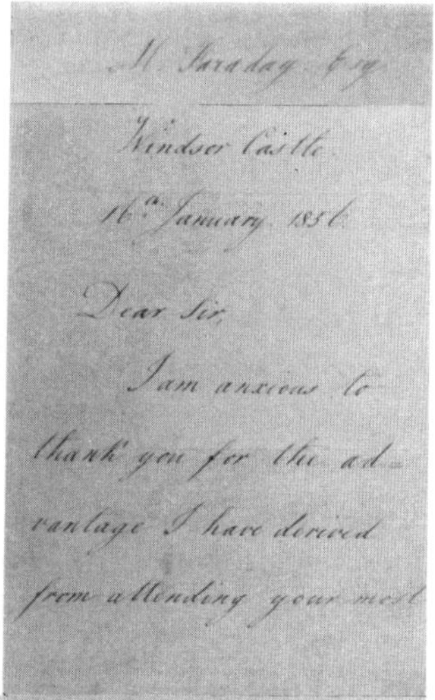

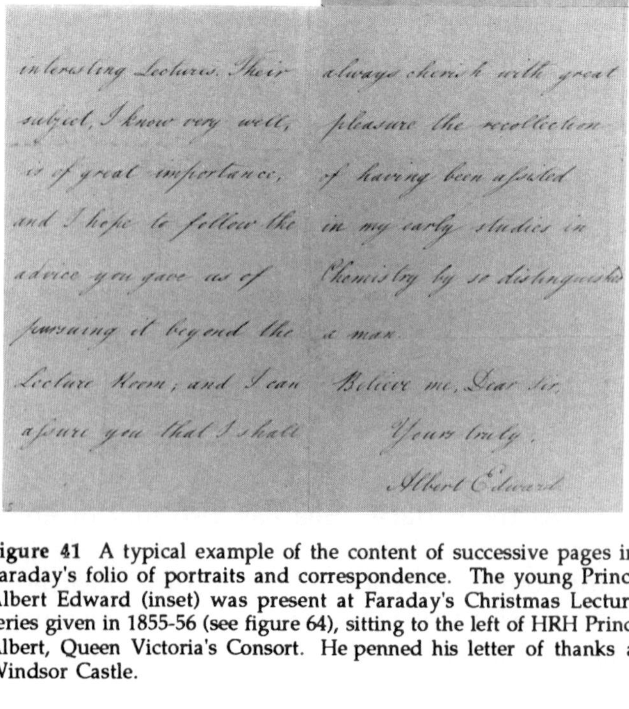

M. Faraday Esq

Windsor Castle

16th January 1856.

Dear Sir,

I am anxious to thank you for the advantage I have derived from attending your most interesting Lectures. Their subject, I know very well, is of great importance, and I hope to follow the advice you gave us of pursuing it beyond the Lecture Room; and I can assure you that I shall always cherish with great pleasure the recollection of having been assisted in my early studies in Chemistry by so distinguished a man.

Believe me, Dear Sir,

Yours truly,

Albert Edward

Figure 41 A typical example of the content of successive pages in Faraday's folio of portraits and correspondence. The young Prince Albert Edward (inset) was present at Faraday's Christmas Lecture series given in 1855-56 (see figure 64), sitting to the left of HRH Prince Albert, Queen Victoria's Consort. He penned his letter of thanks at Windsor Castle.

company of children added an extra resonance to his life, not surprising perhaps in someone whose child-like innocence and perception contributed to his genius.

Much has already been written about the role of religion in Faraday's life.[6] Some insight into the specific religious precepts which he held to be important may be gained from the extensive markings and marginalia in, as well as a few notes and emendations made on, some of Faraday's much-thumbed Bibles. He heavily underlined *Timothy*, **VI**, 10: 'The love of money is the root of all evil'; and the cry of *Job*: 'If I justify myself, my own mouth shall condemn me: if I say I am perfect, it shall also prove me perverse.' But many of his moral principles, to which he adhered with such incorruptible tenacity throughout his life, were of the kind that would appeal to people of goodwill, religious or otherwise, in all ages. Who would dispute the wisdom of the following declaration?

A philosopher should be a man willing to listen to every suggestion but determined to judge for himself. He should not be biased by appearances, have no favourite hypothesis, be of no school and in doctrine have no master. He should not be a respecter of persons, but of things. Truth should be his primary object. If to these qualities be added industry, he may indeed hope to walk within the veil of the temple of nature.

The profound belief — a powerful propellant and guiding star throughout all his scientific quests — in the interconnection and unity of natural forces and phenomena was a conviction that he possessed from an early age. Most religious people in the early 19th century believed in underlying unity. 'God had fashioned the world; it is a whole, so everything should surely be connected', was how the argument went. This unified view of the world was also espoused by many non-religious ancient and more modern philosophers. In his 1821 historical survey (page 29), Faraday drew specific attention to Ørsted's constancy in pursuing inquiries 'respecting the identity of chemical, electrical and magnetic forces.' Even earlier, Davy, in one of the lectures taken down by Faraday in 1812, alluded to a future, simpler science that 'will connect mechanical and chemical sciences.' Faraday was not alone in London's scientific circles in the 1840s in seeking correlations between various natural forces. W R Grove (page 144), the man credited with the enunciation of the First Law of Thermodynamics, pursued this subject with conspicuous success in an extensive course given at the London Institution in 1842. Grove's Bakerian lecture to the Royal Society in 1847 as well as his other publications dwell on this theme. W R Grove's book *Correlation of Physical Forces* was much consulted from the 1850s onwards. In the preface to the last edition of this book (1884), Grove writes:

> *Every one is but a poor judge where he is himself interested, and I therefore write with diffidence; but it would be affecting an indifference which I do not feel if I did not state that I believe myself to have been the first who introduced this subject as a generalized system of philosophy, and continued to enforce it in my lectures and writings for many years, during which it met with the opposition usual and proper to novel ideas.*

Faraday's inner convictions led him to devote most of his energies to two principal activities: pursuing the eternal verities in his role as natural philosopher; and serving his small circle of friends in the Sandemanian sect and the even smaller circle of family members. In the former, his task was not only to prove but to explain: in the latter to offer selfless devotion and compassionate care, which spilled over into work of a charitable kind. He lived a contented, if exhausting, life. And the nature, if not the secret, of that contentedness is revealed in correspondence with his life-long friends — see, especially the letter to De La Rive, page 105. An elaboration of that contentedness as well as his general advice on dealings with other scientists, emerges from his letter to John Tyndall, in connection with an acrimonious discussion at a meeting of the British Association, 6 October 1855:

My dear Tyndall,

These great meetings, of which I think very well altogether, advance science chiefly by bringing scientific men together and making them to know, and be friends with each other, and I am sorry when that is not the effect in every part of their course. I know nothing except from what you tell me, for I have not yet looked at the reports of the proceedings; but let me, as an old man who ought by this time to have profited by experience, say that when I was younger I found I often misintepreted the intentions of people and found that they did not mean what at the time I supposed they meant, and further that as a general rule, it was better to be a little dull of apprehension where phrases seemed to imply pique, and quick in perception when, on the contrary they seemed to imply kindly feeling. The real truth never fails ultimately to appear, and opposing parties, if wrong, are sooner convinced when replied to forebearingly than when overwhelmed. All I mean to say is that it is better to be blind to the results of partisanship and quick to see good will. One has more happiness in oneself in endeavouring to follow the things that make for peace. You can hardly imagine how often I have been heated in private when opposed — as I have thought unjustly, and supercilliously, and yet I have striven, and succeeded I hope, in keeping down replies of a like kind. And I know I have never lost by it. I would not say all this to you did I not esteem you as a true philosopher and friend.

Faraday had little compunction about declining high honours, in turning down invitations to socialize, or in avoiding interruptions. The delegation that implored him to become President of the Royal Society were left in no doubt about his refusal of that honour. Shortly thereafter he remarked:

Tyndall, I must remain plain Michael Faraday to the last; and let me now tell you, that if I accepted the honour which the Royal Society desires to confer upon me, I would not answer for the integrity of my intellect for a single year.

He had earlier declined the Presidency of the Chemical Society, which was formed in his fiftieth year. He hardly ever attended its meetings in nearby Burlington House, not because he was antipathetic to the aspirations of the Society but because he felt disinclined to risk losing yet more time from his beloved laboratory and family. It is not surprising, given these tendencies, that Wheatstone, in a letter to his friend, W F Cooke, dated 4 October 1838, should have said:

I called on Faraday this morning and was told that this was one of the days on which he denies himself to every body for the purpose of pursuing uninterruptedly his own researches. He will be visible tomorrow.

His laboratory hours were long; nine in the morning until eleven at night was not untypical. The only person who ever actively assisted him in his experimental work, and who shared his laboratory, was an ex-Sergeant of the Royal Artillery named Charles Anderson. He had been recruited to maintain the special furnace installed in the Royal Institution for the glass project started in 1827 (see pages 82-83).[7] Hours would pass without a word spoken between them, and when the silence was broken, it was only for a few minutes.

Faraday's letters to the Press, an imperfect measure of one's involvement in the affairs of the day, were very infrequent. Once he wrote to draw attention to the polluted state of the river Thames, an action which prompted *Punch* to publish the cartoon shown in figure 42. Faraday drifted down the Thames inserting strips of cardboard into the water till he could not see the bottom edge and then marking the depth of insertion and the place where the test took place. Another occasion was in 1853 when he was bombarded with requests and enquiries pertaining to the supposed phenomenon of table-turning. This allegedly arose because of the consequences of the spiritualist movement which began in 1848 in Hydesville, New York with the rappings and knockings of the Fox girls.[9] It spread rapidly to other parts of the United States, this country and continental Europe. The spiritual phenomena encompassed the tilting and levitation of tables and chairs, and the movement of objects in the dark became the subjects of several books published in England and France in 1853. The rather rapid and uncritical acceptance of spiritualistic forces greatly disturbed Faraday because of what they revealed about the general level of intelligence.

His exasperation may be gauged from the following excerpt of a letter that he wrote to Professor C F Schönbein, the German-Swiss scientist who discovered ozone. On 25 July 1853 he wrote:

PUNCH, OR THE LONDON CHARIVARI.—July 21, 1855.

FARADAY GIVING HIS CARD TO FATHER THAMES;

And we hope the Dirty Fellow will consult the learned Professor.

Figure 42 The cartoon published in *Punch, or The London Charivari*, 21 July 1855, after Faraday had written to the editor of *The Times* describing how he tested the polluted state of the Thames and pleading that 'surely the river which flows for so many miles through London ought not to be allowed to become a fermenting sewer.'[8]

I have not been at work except in turning the tables upon the table-turners, nor should I have done that, but that I thought it better to stop the inpouring flood by letting all know at once what my views and thoughts were. What a weak, credulous, incredulous, unbelieving, superstitious, bold, frightened, what a ridiculous world ours is, as far as concerns the mind of man. How full of inconsistencies, contradictions, and absurdities it is...

Faraday's letter to *The Times* 'On Table-turning' ended with the sentence:

I think the system of education that could leave the mental condition of the public body in the state in which this subject has found it, must have been greatly deficient in some very important principle.

His concerns about public credulity and general state of awareness, were raised in a famous lecture on 'Mental Education' that he gave at the Royal Institution in May 1854 in the presence of The Prince Consort.

Several scientific anecdotes have centred on Faraday's life and attitudes. Two are very widely quoted, though their authenticity is not unquestioned.[10] Irrespective of whether these anecdotes are true or not, they carry an interesting message. They appear in various forms in the works of 19th and 20th century writers. When Faraday was endeavouring to explain to Prime Minister Robert Peel or to Chancellor of the Exchequer W E Gladstone an important new discovery in science, the politician's alleged comment was 'But, after all, what use is it?' Whereupon Faraday replied 'Why sir, there is the probability that you will soon be able to tax it.' It would be surprising if Faraday did retort in these terms: he seemed to have been singularly uninterested in patenting his own inventions or in the mechanics of wealth creation and taxation. The other retort for which Faraday is often given authorship again involves the Prime Minister or any other earnest enquirer after hearing of Faraday's scientific discovery. The question this time is 'What good is it?' The sagacious reply: 'Of what good is a new-born baby?' is thought to have been first used by Benjamin Franklin in Paris in 1783.

An Analysis of Faraday's Genius

In Owen Meredith's *Last Words of a Sensitive Second-Rate Poet* we read that: 'Genius does what it must, and talent does what it can.' This certainly applies to Faraday just as do other definitions of genius. For example, 'A supreme capacity of taking trouble,' (Samuel Butler); or 'A greater aptitude for patience,' (Comte De Buffon). But none of these goes sufficiently deep. Perhaps Emerson went somewhat deeper if we accept that his words apply more to the realm of natural rather than to moral philosophy: 'To believe your own thoughts, to believe that what is true for you in your private heart is true for all men — that is genius.' But this is still inadequate, so far as Faraday is concerned.

Every great human being of first rank is unique, and Faraday's genius is the consequence of a unique combination of a uniquely large sub-set of major qualities: an infinite capacity to take pains; restless intellectual energy and inexpugnable intellectual honesty, coupled with a measure of technical virtuosity that encompassed the manipulative dexterity and constructive imagination to produce new instruments and new techniques of unsurpassed power and sensitivity. (His torsion balances and coulometers were more sensitive, his electromagnets stronger, his glass specimens heavier and of superior optical quality, than those of his predecessors or contemporaries.) Always convinced that to the problems he pursued there were solutions and that to the questions he raised there were intelligible answers, he had the supreme gift of selecting those that were really important and also of knowing precisely what to do next. Both his strategy and his tactics were impeccable. Add to all this his prodigious physical stamina, endless curiosity, penetrating intuition, complete mastery of detail and an exceptional facility for arguing from the particular to the general — on his own, and with his own brand of self-criticism and self-discipline — and one sees why even those who themselves are regarded as princes among experimenters elevate Faraday to the status of paragon. Rutherford spoke for all scientists when, in 1931, he said:

> *The more we study the work of Faraday with the perspective of time, the more we are impressed by his unrivalled genius as an experimenter and a natural philosopher. When we consider the magnitude and extent of his discoveries and their influence on the progress of science and of industry, there is no honour too great to pay to the memory of Michael Faraday — one of the greatest scientific discoverers of all time.*

Much earlier, John Tyndall in his book *Faraday as Discoverer* analysed the qualities that made Faraday such a successful scientist:

> *He united vast strength with perfect flexibility. ...The intentness of his vision in any direction did not*

apparently diminish his power of perception in other directions, and when he attacked a subject expecting results, he had the faculty of keeping his mind alert, so that results different from those he expected should not escape him through preoccupation.

To all this analysis must be added three other factors. Firstly, Faraday wrote and spoke about his work in memorable ways. Secondly, almost all the successful experiments that he carried out he proceeded to refine, with a view to demonstrating them publicly at Discourses in the Royal Institution. They were intended to leave an indelible impression and in this he succeeded triumphantly. Lastly, he had the extra good fortune to have as one of his interpreters arguably one of the greatest physicists since Newton, J Clerk Maxwell. Maxwell, who was born the year that Faraday made his most momentous discovery (of electromagnetic induction) in 1831, selected 'Faraday's Lines of Force' as the title of his extraordinary paper delivered to the Cambridge Philosophical Society in December 1855 and February 1856, when he was a twenty-four year old Fellow at Trinity College, Cambridge. With that monumental work[11] mathematical precision and quantitative prediction were added to Faraday's qualitative views on field theory in general and to electromagnetism in particular. With this event a new era dawned.

Notes

1. Angela Burdett-Coutts, a member of the banking family, was persuaded by Faraday to join the Royal Institution. They were both patrons at the Orphan Asylum in the 1840s. On 29 May 1856 Faraday

joined guests at the roof of the Baroness's home to celebrate the end of the Crimean War. It is reported (see Edna Healey's *Lady Unknown* (Sigwick Jackson, London) 1978, 62) that, on watching the fireworks, Faraday 'hallooed out with wonderful vivacity, "there goes magnesium, there's potassium."'

2. An editorial published in *The Referee*, 21 June 1891, discloses that

> *Michael Faraday was one of the elders of our chapel; another was a butcher, another a gas fitter, and a fourth, if I remember rightly, a linen draper. I heard Faraday read the Bible & expound often during my childhood, and I remember I liked him best of all the elders because he didn't waggle his head and whine & tremble like some of the others.*

3. *Proc. of the Royal Inst. of G.B.* **V**, 214 (1868).

4. Some of it is still in use in the Director's second floor flat and in offices and public sectors of the Royal Institution.

5. His physical strength never deserted him. As adviser to Trinity House, he was called upon to make visits to various lighthouses. At the age of seventy he braved snow and storm, crossed fields and hedges, and put up with other discomforts in his tour of duty.

6. See for example H Marryat, *The Times*, 21 September 1931; G A Cantor, *B.J.H.S.*, **22**, 433 (1989); H T Pratt, Amer. Chem. Soc. Meeting, Atlanta, April 1991 (abstract).

7.   From 1830 to 1833 Faraday paid Anderson's salary from his own pocket. Thereafter, up until the time of his death nearly twenty years later, the Royal Institution paid for Anderson as an assistant.

8.   Faraday's letter to the Editor of *The Times*, 7 July 1855, reads as follows:

> *Sir, — I traversed this day by steamboat the space between London and Hungerford Bridges, between half-past one and two o'clock. It was low water, and I think the tide must have been near the turn. The appearance and smell of the water forced themselves at once on my attention. The whole of the river was an opaque pale brown fluid. In order to test the degree of opacity, I tore up some white cards into pieces, and then moistened them, so as to make them sink easily below the surface, and then dropped some of these pieces into the water at every pier the boat came to. Before they had sunk an inch below the surface they were undistinguishable, though the sun shone brightly at the time, and when the pieces fell edgeways the lower part was hidden from sight before the upper part was under water.*
>
> *This happened at St Paul's Wharf, Blackfriars Bridge, Temple Wharf, Southwark Bridge, and Hungerford, and I have no doubt would have occurred further up and down the river. Near the bridges the feculence rolled up in clouds so dense that they were visible at the surface even in water of this kind.*
>
> *The smell was very bad, and common to the whole of the water. It was the same as that which now comes up from the gully holes in the streets. The whole river was for the time a real sewer. Having just returned from the country air, I was*

perhaps more affected by it than others; but I do not think that I could have gone on to Lambeth or Chelsea, and I was glad to enter the streets for an atmosphere which, except near the sink-holes, I found much sweeter than on the river.

I have thought it a duty to record these facts, that they may be brought to the attention of those who exercise power, or have responsibility in relation to the condition of our river. There is nothing figurative in the words I have employed, or any approach to exaggeration. They are the simple truth.

If there be sufficient authority to remove a putrescent pond from the neighbourhood of a few simple dwellings, surely the river which flows for so many miles through London ought not to be allowed to become a fermenting sewer. The condition in which I saw the Thames may perhaps be considered as exceptional, but it ought to be an impossible state; instead of which, I fear it is rapidly becoming the general condition. If we neglect this subject, we cannot expect to do so with impunity; nor ought we to be surprised if, ere many years are over, a season give us sad proof of the folly of our carelessness.

9. See *Scientific Monthly*, September 1956, 145.

10. See C J Webb and I B Cohen, *Nature*, **157**, 196 (1946); and R A Gregory, *Nature*, **157**, 305 (1946).

11. *Proc. Camb. Phil. Soc.*, **X**, Part I, (1856).

Chapter 7

Faraday's Influence upon The Royal Institution

At 9pm on the evening of 24 February 1978, the distinguished Vienna-born art historian, Sir Ernst Gombrich began a lecture entitled 'Experiment and Experience in the Arts' with the following words:

It seems to me a pleasant fancy to imagine that no event in this world ever disappears without trace, and that even the words spoken in a particular room continue to reverberate, ever more slightly, long after their audible echoes have faded. If that were true a supersensitive instrument might still be able to pick up the resonance of words spoken in this very hall a little less than 142 years ago in what I imagine to have been a vigorous Suffolk accent, very different from mine. 'Painting is a science' — you would hear the voice say — 'and should be pursued as an inquiry into the laws of Nature. Why, then, may not landscape painting be considered as a branch of natural philosophy, of which pictures are but the experiments'? The artist who was thus appealing to the genius loci of this place was John Constable and the occasion of the last of four lectures he gave at the Royal

134

Institution in April 1836, for which the invitation is still preserved in its library.

𝕽𝖔𝖞𝖆𝖑 𝕴𝖓𝖘𝖙𝖎𝖙𝖚𝖙𝖎𝖔𝖓 ⟨ 𝕲𝖗𝖊𝖆𝖙 𝕭𝖗𝖎𝖙𝖆𝖎𝖓,
ALBEMARLE STREET.

23d April, 1836.

SYLLABUS
OF A COURSE OF LECTURES
ON THE
HISTORY OF LANDSCAPE PAINTING,
BY
JOHN CONSTABLE, Esq. R. A.

To be delivered on the following Thursdays at Three o'Clock.

LECTURE I. May 26.—The real Origin of Landscape—Coeval in Italy and Germany in its rise and Early progress—Further Advanced in Germany in the Fifteenth Century—Albert Durer—Influence of his Works in Italy—Titian—impressed by them and in *his* hands Landscape assumed its real dignity and grandeur—and entitled him to the appellation of the " *Father of Landscape*"—the " St. Peter Martyr."

LECTURE II. June 2.—Establishment of Landscape—the Bolognese School —by this School Landscape first made a distinct and separate Class of Art—the Sixteenth and Seventeenth Centuries — the Caracci — Domenichino — Albano—Mola—Landscape soon after perfected in Rome—the Poussins — Claude Lorraine — Bourdon — Salvator Rosa — The " Bambocciate " — Peter de Laar — Both —Berghem —the deterioration of Landscape—its Decline in the Eighteenth Century.

LECTURE III. June 9.—Landscape of the Dutch and Flemish Schools—emanates from the School of Albert Durer—forming separate and distinct branches—Rubens—Rembrandt—Ruysdaal—Cuyp—the marks which characterize the two schools—their decline also in the Eighteenth Century.

LECTURE IV. June 16.—The decline and revival of Art—imitation of preceding excellence the 'main cause of the decline—opposed to original Study—the Restoration of Painting takes place in England—Reynolds—Hogarth—West—Wilson—Gainsborough —when Landscape at length resumes its birthright—and appears with new powers.

Gentlemen as well as Ladies are admitted as Subscribers to the Lectures on payment of Two Guineas for the Season, or One Guinea each Course.

London : William Nicol, Printer to the Royal Institution.

Figure 43 Content of course of lectures given by the landscape painter John Constable at the Royal Institution in 1836.

Gombrich's evocative opening reminds us that almost every living scientist could, with appropriate substitution of the proper nouns, begin his or her Friday Evening Discourse in a plagiarized form of the above passage: astronomers, botanists, chemists, dermatologists right through to zoologists could quote the night on which their heroes or the founder and protagonists of their subject spoke at the Royal Institution, for almost every major scientist of the 19th century, and a goodly fraction of the giants of the first half of the 20th century, performed here, and left a lasting impression. The list of performers also extends to artists, architects, explorers, lawyers, musicians and parliamentarians. Speakers who are nowadays invited to participate in contemporary programmes of Discourses have been known to brood upon, if not to be daunted or inspired by, a seemingly endless succession of glittering predecessors. These include the first Swedish winner of the Nobel prize, Svante Arrhenius (1857-1927) who, in 1911, considered 'Applications' of physical chemistry to the doctrine of immunity, antigens and antibodies'; the Astronomer Royal of his day and President of the Royal Society, Sir George Airy, who, in 1851 described the total solar eclipse of July that year; Alfred Austin, the legally-trained journalist who succeeded Tennyson as Poet Laureate and who spoke in 1904 on 'The growing distaste for the higher kinds of poetry'; the critic Matthew Arnold (1822-1888) who drew a massive audience (868) in March 1884 when he spoke on 'Emerson'; and the pioneer physicist F W Aston (1877-1945) who in 1921 showed how elegantly his mass spectrometer could identify the isotopes of gaseous elements such as argon and neon. Proceeding to the second letter of the alphabet we note that the resourceful Indian comparative physiologist, Jagadis Chunder Bose (1858-1937), dealt with 'Plant-autographs and their metabolism' in 1914; that the colourful explorer Samuel White Baker (1821-1893) discussed the 'Source of the Nile' which he did much to discover,[1] in 1866; that the geneticist William Bateson (1861-1926) who became the first director of the John Innes Horticultural Institute, and who was instrumental in retrieving from potential oblivion Mendel's seminal work, lectured on 'Fifteen years of Mendelism' in 1916; that Henri Becquerel described, in 1902 (in French), his discovery of radioactivity; that the Scottish-born American inventor Alexander Graham Bell (1847-1922) discoursed on 'Speech' in 1878; that the choral-singing English physicist Nobel prizewinner Charles Barkla spoke on X-rays in 1916; and that between them Sir William and Sir Lawrence Bragg in the period 1911 to the late 1960s in a scintillating sequence of some fifty Discourses covered most topics in the science of the crystalline state and much else.

The Royal Institution has played a historic role in British scientific and cultural life for nearly two centuries. For more than a quarter of that time its efforts and policies reflected the genius of Michael Faraday. From its inception it has remained independent and does not receive any direct government funding. Its success and very existence rely on the support given to its resident scientists and staff by its members and the numerous private organizations interested in the discovery and dissemination of scientific knowledge. It is indisputably the most famous 'theatre' of science in the world, thanks largely to the personality of Humphry Davy (its brilliant Director shortly after its birth), and especially to traditions initiated and perfected by

Faraday and devotedly sustained by his eminent successors and supporters.

A measure of the range of subjects that Faraday brought to the attention of his audiences is gained from the tabulations of topics covered both by him and by those whom he invited to speak at Friday evening Discourses (see Appendices I, III and IV). The subjects selected by Faraday during his last twenty seven years as Director (see Appendix IV) went much beyond the realms of his own scientific research interests (compare with Appendices I and III); and, judging by the attendance figures, his popularity as a speaker reached a level which has never since been surpassed.

Faraday himself set the tone of the Friday Evening Discourses and the time-honoured practices that evolved during the latter half of the 19th century are still maintained. Promptly at nine o'clock, the speaker, having been released from his incarceration a few minutes earlier, enters the auditorium unannounced. The audience is formally dressed as at an operatic performance, and after exactly an hour's discourse, the speaker is led out by the Director. Members and guests converse during the serving of refreshments and may view the range of special exhibits (on the theme of the discourse) in the beautiful library — no longer two-tiered as in the print shown in figure 5 but equally inviting and filled with a profound sense of history.[2] They may also view Faraday's museum containing his equipment, samples, original notebooks and medals, or saunter through other quarters rich in portraits and relics of earlier eras.

As the titles of the talks given in 1832-62 reveal (see Appendices I, III and IV), it was Faraday's custom to cover a wide diversity of topics in each season's set of Discourses. Faraday himself embraced many subjects, sometimes advised and guided by his close associates, notably his friends Hulmandell, the lithographer, and Wheatstone who, being shy (see page 72), preferred that Faraday should describe his discoveries and inventions, rather than he himself. Faraday also persuaded many of his eminent contemporaries to present Discourses. For example, George Gabriel Stokes (1819-1903), the Irish-born, Cambridge natural philosopher and Lucasian Professor of Mathematics there, gave, in February 1853, the first ever public demonstation of the phenomenon of fluorescence, which he himself discovered. A little later, Warren De La Rue (1815-1889), the pioneer astronomical photographer, showed the results taken during the Spanish eclipse expedition (1860) which proved that the red flames known as 'prominences' were of solar not lunar origin. Faraday himself gave one of the earliest public demonstrations of photography at a Discourse with the cooperation of Fox Talbot. Indeed, the first recorded electrical flash photograph was taken in the lecture theatre that night. W H Fox Talbot (1800-77), one of the founders of photography, carried out some of his work at the Royal Institution. He ends a letter to Faraday, dated 15 June 1851, thus:

If a truly instantaneous photographic representation of an object has never been obtained before (as I imagine it has not) I am glad that it should have been first accomplished at the Royal Institution.

Nowhere in the Royal Institution (which is nowadays part university, part museum, part research centre, part classroom, part library, part club, part exhibition and broadcasting centre amongst other things), are members and visitors more keenly conscious of the continuity of life and the genius of man and place than in the main lecture theatre [3] where Faraday performed on more than a thousand occasions, the scene of countless elegant demonstrations and memorable expositions by him and others. A pioneering example of the use of the lecture theatre at the Royal Institution as a broadcasting studio occurred at midnight 26-27 January 1987 when the Research Director of the British Petroleum Company, Professor (now Sir) John Cadogan presented a lecture-demonstration via satellite to the Annual Meeting of the Australia and New Zealand Association for the Advancement of Science at Palmerston, New Zealand. An invited audience of 200 watched him perform at the Royal Institution on the theme of 'Aspects of an Oil Giant's Research: From Pure Science to Profit' and there were 500 viewers in New Zealand.[4]

But it is not only the popularity or frequency of Faraday's brilliant lectures at the Royal Institution that induces humility among succeeding performers; it is also the thoroughness with which Faraday checked and counter-checked his experiments before they were exposed to the public glare. Sir George (later Lord) Porter, one of the world's most accomplished lecture-demonstrators, tells an interesting tale that illustrates this fact.

Shortly after I came to the Royal Institution to occupy the Chair of Chemistry, which was originally founded for Faraday, I gave a Schools Lecture on the subject of the chemical bond. In order to demonstrate standing wave patterns similar to those of electrons in atoms I used the well-known Chladni plate on which sand is made to fall into the nodes when the plate is set into oscillation by bowing it. Our first violin, Mr Coates here, will illustrate these nodal patterns. [See figure 44.]

(a)

(b)

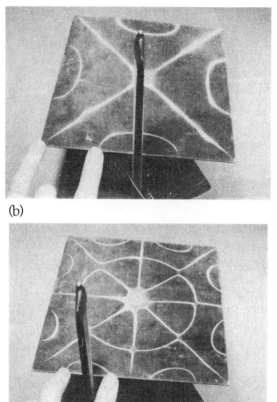

Figure 44 When sand is sprinkled on a thin metal (Chladni) plate and the latter is set vibrating by touching a side with a 'sonorous' violin bow, beautifully symmetric acoustic patterns (studied by Faraday in 1831) are produced. The frequencies used in (a) and (b) are respectively 1200 and 3828 cycles per second (after Charles Taylor, *The Art of Science of Lecture Demonstration*, Adam Hilger, 1988).

Figure 45 William Henry Smyth, friend of W R Grove and father of Charles Piazzi Smyth. Each letter he wrote to his friend carried a suitable cartoon. (See figure 50.) (From *The Peripatetic Astronomer: The Life of Charles Piazzi Smyth* by H A Brück and M T Brück (Adam Hilger, Bristol, 1988).)

Now, something quite unexpected happened. We ran out of sand and Mr Coates gave to me some talcum powder instead, on the reasonable expectation that it would behave similarly. It did not behave similarly, in fact it behaved in quite the opposite manner and the powder occupied the antinodes, where the oscillation was greatest — as you see, it falls into the spaces between the lines of sand. I invited my colleague, Professor King, to witness this remarkable phenomenon but he was neither impressed nor surprised because, he told me, a full account and explanation of the effect had been published over a century ago. The author? Yes — Michael

Faraday, in the <u>Philosophical Transactions of the Royal Society</u>, 1831. In that paper he described 129 experiments covering all aspects of these vibrating plates and he showed that, whilst the heavier sand particles fall into the nodes, the lighter powder is drawn into the antinodes by the current of air which is created as the plate rises. Being Faraday, he provided the definitive proof by performing the experiment in a vacuum and showing that the light powder then behaved like the heavier sand.

Of the numerous individuals whom Faraday invited to present Discourses in his day two merit especial attention: Charles Piazzi Smyth (1819-1900) and William Robert Grove (1811-1896). These were unconventional men, the latter being a close friend of the former's flamboyant father William Henry Smyth.

Charles Piazzi Smyth, who was for forty years the Astronomer Royal for Scotland, was one of the most colourful personalities in the world of British science in the 19th century. Apart from being an astronomer and a spectroscopist, he was also a pioneer photographer, a meteorologist, a metrologist, artist, traveller, writer and pyramidalogist, believing that there was some mystical significance to the dimensions of the Great Pyramid of Gizeh in Egypt.

Piazzi Smyth made one contribution of paramount importance to his subject, and that was his advocacy of 'mountain astronomy'. His experiment in 1856, about which he spoke in his Royal Institution Discourse on 5 March 1857, on the Peak of Tenerife demonstrated that there — to borrow Isaac Newton's words — in 'serene air above the grosser clouds' lay the future for observational astronomy. The standard practice of optical and other astronomers of placing their observatories at high altitudes, a practice which is nowadays pursued worldwide and leads to the steady

Figure 46 Charles Piazzi Smyth, the first of the peripatetic astronomers. (From *The Peripatetic Astronomer: The Life of Charles Piazzi Smyth* by H A Brück and M T Brück (Adam Hilger, Bristol, 1988).)

migration of astronomers to the attractive climes of California, the Canary Islands and Hawaii, dates from Piazzi Smyth's original pathfinding expedition to Tenerife. He was the first of the peripatetic astronomers.[5]

William Robert Grove (see page 122) was an eminent lawyer and man of science who died in his 86th year in 1896. He was the son of John Grove, a magistrate and deputy-lieutenant of the county of Glamorgan and was born at Swansea. He received his early education from the Rev Eli Griffiths before proceeding at a tender age to Brasenose College, Oxford, where he took an ordinary degree in 1830. He was called to the Bar at

Figure 47 *SPY*'s cartoon of the scientist-turned judge, William Robert Grove, the man accredited with enunciating the First Law of Thermodynamics. In satirical mood, this cartoon, which was published in *Vanity Fair* was labelled 'Galvanic Electricity'.

Lincoln's Inn in 1835, but was prevented by ill-health for several years from actively pursuing his profession. During this period he was absorbed in electrical researches, and suceeded in 1839 in contriving the powerful battery which bears his name (figure 48). In

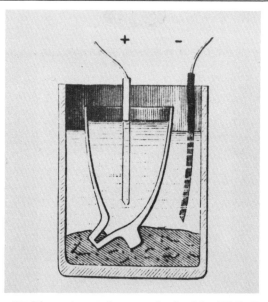

Figure 48 The primary battery devised by W R Grove in 1839 consisted of a number of small cells contained in glass vessels, the electrolytes (conducting solutions) being separated by means of the broken-off bowls of clay tobacco pipes. The positive pole was of zinc dipping into dilute sulphuric acid, and the negative pole of platinum, dipping into concentrated nitric acid. This battery yields nearly two volts. This is what Faraday and others meant when they spoke of Grove's intensity battery.

the following year he was appointed Professor of Experimental Philosophy at the London Institution, a body which no longer exists. (Its original building is now used by the London School of Oriental and African Studies.) It was during his seven-year tenure at the London Institution that he laid the foundation of the European reputation in the realm of physical investigation which he subsequently acquired. His 1842 series of lectures dealt with his pioneering views (congruent with those of Faraday) on the mutual convertibility of the various natural forces. This was when he enunciated the concept of the conservation of energy (the First Law of Thermodynamics). His epoch-making essay on the 'Correlation of Physical Forces' was published in 1846. He gave the Bakerian lecture of the Royal Society in 1847 on 'Voltaic Ignition' and on the decomposition of water into its constituent gases by a heated platinum wire, thereby greatly clarifying the nature of the phenomenon of catalysis. (A heated platinum wire also facilitates the co-elimination of hydrogen and oxygen to form water. A catalyst, therefore, merely assists in the attainment of equilibrium.)

The Grove cell (figure 48) is not to be confused with his equally famous 'gas battery', or fuel cell, which he was the first to invent. He described it in a charming letter to Michael Faraday written on Saturday October 22, 1842 (figure 49). It reads:

> *W.R. GROVE TO M. FARADAY*
> *London Institution*
> *Saturday Oct 22 1842*

My dear Sir

I have just completed a curious voltaic pile which I think you would like to see, it is composed of alternate tubes of oxygen & hydrogen through each of which passes platina foil so as to dip into separate vessels of water acidulated with sulphuric acid the liquid just touching the extremities of the foil as in the rough figure below.

The platina is platinized so as to expose by capillary attraction a greater surface of [liquid] to the gas, with 60 of these alternations I get an unpleasant shock & decompose not only iodide of potassium but water so plainly that a continuous stream of thin bubbles ascends

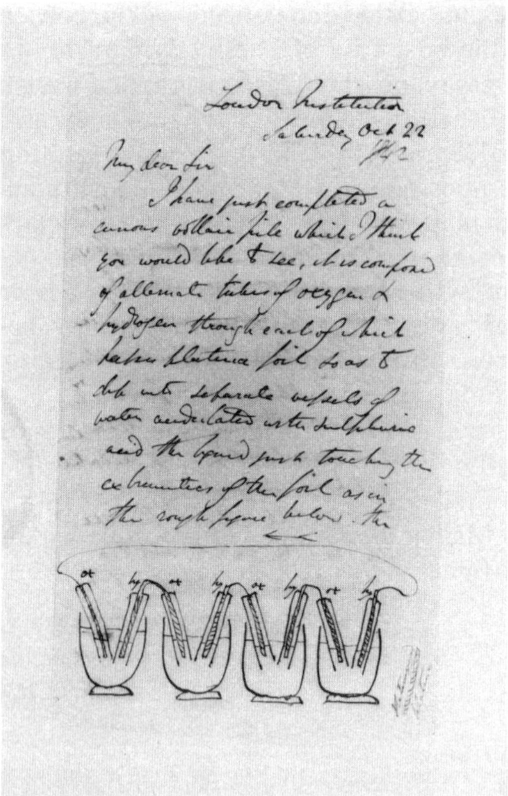

Figure 49 The famous letter written by Grove to Faraday in October 1842. This contains a description of what Grove called his 'gas battery' which is, in fact, the first fuel cell. Hydrogen and oxygen are 'burned' slowly by electrochemical action, thereby generating a voltage. (Banks of (Grove) fuel cells based on this principle are now of great practical importance in generating electricity directly from chemical reactions. Transport fuel in future may operate on this principle; breathalysers (for alcohol) already do.)

from each electrode Thus water is decomposed by its composition — no oxidable metal is employed. I have reversed the tubes & tried all the counter expts but the phenomena are too marked I think to render any mistakes possible. Mr Gassiot was with me today & saw the Expts. Can you spare me an hour next week on Tuesday if it suits you or any day except Wednesday at any hour from 11 to 3 — at the Laboratory of the London Institution. I cannot but regard the experiment as an important one both as to the chemical & other theories of the pile & as to the catalytic effects of the combination of the gases by platina.

I remain my dear Sir
yours very sincerely
W.R. Grove

At his Royal Society lecture given on 11 May 1843 Grove described further characteristics of his 'curious Voltaic pile' (or gas Voltaic battery).

The platinum in the oxygen of one pair was metallically connected with the platinum in the hydrogen of the next, and a voltaic series of fifty pairs was thus formed. With this battery the following effects were produced:
A shock was given which could be felt by five persons joining hands; and a brilliant spark visible in broad day-light was given between charcoal points.

Grove was rather fond of conducting spectacular, even shocking, demonstrations. During the course of the first of his fourteen Discourses at the Royal Institution, given on 13 March 1840 Faraday, who volunteered some experimental assistance that night, came near to burning his fingers! Thus we read in the April 1840 issue of *The Annals of Electricity, Magnetism and Chemistry;* and *Guardian of Experimental Science,* a journal edited by William Sturgeon:

Mr. Grove then referred to his 'intensity' battery which only covered a square surface of 16 inches on each side. The cells were 4 inches high and consisted of 50 pairs of platinum plates 2 inch x 3 inch, with double amalgamated zincs. With this comparatively small intensity-battery an arc of flame, between charcoal points, was observed, 1 $1/4$ inch long, fuse large and thick iron wire. Mr. Faraday having lent his pocket knife to the lecturer, the large blade was instantly deflagrated, exhibiting a splendid shower of scintillations of steel. Large masses of zinc, copper, soft iron, &c. were then submitted to the action of the battery, and a most splendid series of combustions were the consequence, the colour of the light being dependent upon the metal employed.

Sturgeon (1783-1850) was an electrician and inventor. Born in humble circumstances Sturgeon first became a private soldier and later educated himself in languages and natural science. He left the Army in 1820, set himself up as a boatmaker in Woolwich, where he later met Faraday. In 1824 he became a Lecturer in Science at the Royal Military College. He later moved to Manchester, but in the meantime he was among the first to build a workable electric motor. Grove returned to the practice of law and became a Queen's Council in 1853, and a judge in 1871. His special knowledge of science serving him well in cases involving infringement of patents. He was one of the original members of the Chemical Society, and a Vice-President of the Royal Society and of the Royal Institution. He was also a very keen amateur astronomer, and a strong supporter of the British Association of which he was President at its Nottingham meeting in 1866.

In March 1891, Grove made a memorable speech at the Chemical Society's jubilee, at which he mentioned that he was one of the original members of the society, and was present at the lecture given by John Dalton at the Royal Institution on 9 May 1834 when the latter lectured 'On the Atomic Theory of Vapours.' During his speech he also reproached himself for not having discovered the spectroscope.

> *I had obserbed that there were different lines exhibited in the spectra of different metals ignited in the Voltaic arc, and if I had had any reasonable amount of wit I ought to have seen the converse — viz., that by ignition different bodies show in their spectral lines the materials of which they are composed.*

Had he exploited this realization earlier he would have been ahead of the great German chemists Bunsen and Kirchhoff, commonly regarded as progenitors of modern spectroscopy. At the same jubilee meeting Grove said: 'For my part, I must say that science to me generally ceases to be interesting as it becomes useful.' There is some irony in the fact that much of his work led to important practical consequences. Yet his contribution to the concept of energy conservation, of which he was justifiably proud (page 147), was overshadowed by the work of others including Faraday's.

Grove and Piazzi Smyth's father W H Smyth were devoted friends, brought together by a common love for astronomy and the extra attractions of the Royal Astronomical Society's dining club. They regularly dined together after meetings. They also frequently retreated to Hartwell House in Buckinghamshire, where, under Smyth's guidance, a proper astronomical observatory had been erected. The archives of the Royal Institution contain many of the original letters received by Grove from W H Smyth, who had the endearing habit of including a cartoon sketch depicting his mood at the moment of writing. Smyth's style is characteristic of the era (figure 50). The archives also include a rich collection of the letters received by Grove from Charles Wheatstone and W H Fox Talbot (see page 139).

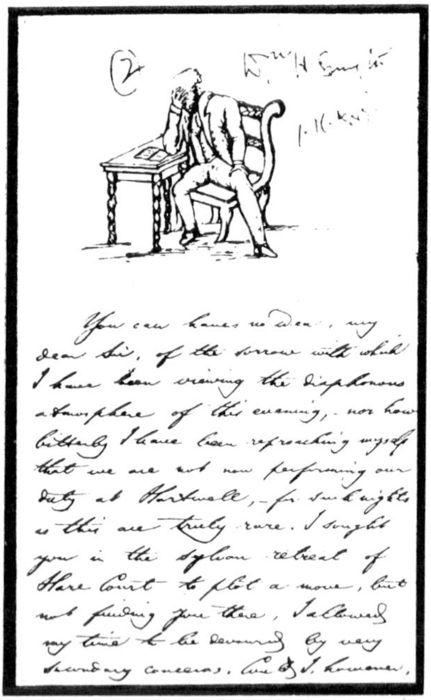
You can have no idea, my dear Sir, of the sorrow with which I have been viewing the diaphanous atmosphere of this evening, - nor how bitterly I have been reproaching myself that we are not now performing our duty at Hartwell, - for such nights as this are truly rare. I sought you in the sylvan retreat of Hare Court to plot a move, but not finding you there, I allowed my time to be devoured by very secondary concerns.

Figure 50 One of W H Smyth's letters to W R Grove, bemoaning the loss of a perfect evening for astronomical recreation (from the archives of the Royal Institution). The letter reads:
You can have no idea, my dear Sir, of the sorrow with which I have been viewing the diaphanous atmosphere this evening, nor how bitterly I have been reproaching myself that we are not now performing our duty at Hartwell, for such nights as this are truly rare. I sought you in the sylvan retreat of Hare Court to plot a move, but not finding you there, I allowed my time to be devoured by very secondary concerns.

Writing to Grove from the Hotel de l'Europe in Venice on 17 July 1868, Fox Talbot alludes to the reasons why he declined the Presidency of the British Association for 1869.

I expressed to him my thanks for the great honour done to me by the Society in selecting my name for the presidency in 1869, and my regret that I cannot possibly accept it. In fact I have no idea whether I shall be in England at that time, the precarious health of members of my family requiring a residence abroad and though I hope to visit England occasionally, yet I am liable to be unexpectedly called away. I therefore cannot undertake any engagement of consequence.

Maintaining the Faradaic Tradition

John Tyndall (1820-1893), a remarkable Irishman, was brought to the Royal Institution by its percipient secretary, the physician Henry Bence Jones (1813-1873), who is now remembered through the proteins which bear his name, and for *The Life and Letters of Faraday* which he edited in 1870. On returning to Britain after a successful period of study under Bunsen at the University of Marburg in Germany, Tyndall was soon appointed (1853) Professor of Natural Philosophy at the Royal Institution. Before long the ageing Faraday and he were established as close colleagues.

Tyndall became an outstanding scientist and maintained the best traditions of the Institution as a centre for research: he explained the flow of glaciers and was the first to measure the absorption and radiation of heat by gases and vapours, and the first to identify the 'greenhouse effect'. He also carried out pioneering work in the scattering of light by small particles suspended in the air, and in this field is remembered by the Tyndall effect. Like Pasteur, Tyndall did much to extinguish the smouldering controversy relating to spontaneous creation — the doctrine that living creatures arise, not by normal reproductive processes from parents like themselves, but *de novo* from movements and perturbations occurring in organic matter which is undergoing decomposition. Tyndall's work on dust and bacteria in the air, some of which he carried out in the Botanical Gardens at Kew as well as on the roof of the Institution, led to the new method of so-called Tyndallization. This is the process of discontinuous sterilization by heating to moderate temperatures: it proved of great value in the early days of bacteriology. Tyndall was also a compelling lecturer who soon won the hearts of the audiences at the Friday Evening Discourses. The range of his subjects maintained the fine traditions initiated by Faraday.

Tyndall's flair for showmanship often came to the fore in his lectures. He once stood under artificially produced showers of water, paraffin, turpentine and petroleum wearing a cape and sou'wester in order to show how raindrops are created. On another occasion he proved the potential intensity of invisible heat emitted from a source by lighting a cigar at the invisible focal point.

He published a large number of books, based mainly on his lectures and researches but aimed at a wider, less sophisticated public than the distinguished audiences that flocked to hear him at the Royal Institution. Some of his books became best-sellers; and he was occasionally taunted for being more of a popularizer than a serious scientist. But most Victorians appreciated his proselytizing zeal for the spread of scientific education.

Tyndall's felicity of expression still elicits admiration. This is how in May 1883 he described to his Discourse audience the slings and arrows directed at Rumford:

> *Rumford and his Institution had to bear the brunt of ridicule, and he felt it; but men of ready wit have not abstained from exercising it on societies of greater age and higher claims. Shafts of sarcasm without number have been launched at the Royal Society. It is perfectly natural for persons who have little taste for scientific inquiry and less knowledge of the methods of nature, to feel amused, if not scandalised, by the apparently insignificant subjects which sometimes occupy the scientific mind. They are not aware that in science the most stupendous phenomena often find their suggestion and interpretation in the most minute, — that the smallest laboratory fact is connected by indissoluble ties with the grandest operations of nature. Thus from the iridescences of the common soap-bubble, subjected to scientific analysis, have emerged in the conclusion that stellar space is a plenum filled with a material substance, capable of transmitting motion with a rapidity which would girdle the equatorial earth eight times a second; while the tremors of this substance, in one form, constitute what we call light, and, in all forms, constitute what we call radiant heat. Not seeing this connection between great and small; not discerning that as regards the illustration of physical principles there is no great and no small, the wits, considering the small contemptible, permitted sarcasm to flow accordingly. But these things have passed away, otherwise it would not be superfluous to remind this audience, as a case in point, that the splendour which in the form of the electric light now falls upon our squares and thoroughfares, has its germ and ancestry in a spark so*

feeble as to be scarcely visible when first revealed within the walls of this Institution.

His skill in capturing the interest of his listeners is well illustrated by the opening paragraphs of his famous Discourse, in 1867, 'On Sounding and Sensitive Flames':

> *The sounding of a hydrogen flame when enclosed within a glass tube was, I believe, first noticed by Dr. Higgins, in 1777. The subject has been since investigated by Chladni, De la Rive, Faraday, Wheatstone, Riyke, Sondhauss, and Kundt. The action of unisonant sounds on flames enclosed in tubes has been investigated by Count Schaffgotsch and myself. The jumping of a naked fish-tail flame, in response to musical sounds, was first noticed by Professor Leconte at a musical party in the United States. He made the important observation that the flame did not jump until it was near <u>flaring.</u> That his discovery was not further followed up by this learned investigator was probably due to too great a stretch of courtesy on his part towards myself. Last year, while preparing the experiments for one of my 'Juvenile Lectures,' my late assistant, Mr. Barrett, observed the effect independently; and he afterwards succeeded in illustrating it by some very striking experiments. With a view to the present discourse, and also to the requirements of a forthcoming work on Sound, the subject of sounding and sensitive flames has been recently submitted to examination in the Laboratory of the Royal Institution. The principal results of the inquiry are embodied in the following abstract.*
>
> *Pass a steadily-burning candle rapidly through the air, you obtain an indented band of light, while an almost musical sound heard at the same time announces the rhythmic character of the motion. If, on the other hand, you blow against a candle-flame, the fluttering noise produced indicates a rhythmic action.*

When a fluttering of the air is produced at the embouchure of an organ-pipe, the resonance of the pipe reinforces that particular pulse of the flutter whose period of vibration coincides with its own, and raises it to a musical sound.

When a gas-flame is introduced into a open tube of suitable length and width, the current of air passing over the flame produces such a flutter, which the resonance of the tube exalts to a musical sound.

Introducing a gas-flame into this tin tube three feet long, we obtain a rich musical note; introducing it into a tube six feet long, we obtain a note an octave deeper — the pitch of the note depending on the length of the tube. Introducing the flame into this third tube, which is fifteen feet long, the sound assumes extraordinary intensity. The vibrations which produce it are sufficiently powerful to shake the pillars, floor, seats, gallery, and the five or six hundred people who occupy the seats and gallery. The flame is sometimes extinguished by its own violence, and ends its peal by an explosion as loud as a pistol shot.

The roar of the flame in a chimney is of this character: it is a rude attempt at music.

By varying the size of the flame, these tubes may be caused to emit their harmonic sounds.

One of the greatest services that Tyndall rendered to the Royal Institution and to posterity is the analysis he gave in succeeding Discourses, on 17 and 24 January 1868, of *Faraday as a Discoverer*. The Proceedings of the Royal Institution for that year contain the full sweep of Tyndall's endeavours. The evocative and winning ways of Tyndall are seen perhaps at their best when he analysed in these Discourses Faraday's approach to the debate concerning the 'Origin of Power in the Voltaic Pile'. The question at issue was whether mere contact of dissimilar metals or other substances was a necessary and sufficient condition to generate electricity, or was it

the consequence of chemical reaction? This is how
Tyndall sets the scene and drives home the message.

*In one of the public areas of the town of Como stands a
statue, with no inscription on its pedestal save that of a
single name, 'Volta.' The bearer of that name occupies a
place for ever memorable in the history of science. To
him we owe the discovery of the voltaic pile, to which,
for a brief interval, we must now turn our attention.*

*The objects of scientific thought being the passionless
laws and phenomena of external nature, one might
suppose that their investigation and discussion would be
completely withdrawn from the region of the feelings,
and pursued by the cold dry light of the intellect alone.
This, however, is not always the case. Man carries his
heart with him into all his works. You cannot separate
the moral and emotional from the intellectual; and thus
it is that the discussion of a point of science may rise to
the heat of a battle-field. The fight between the rival
optical theories of Emission and Undulation was of this
fierce character; and scarcely less fierce for many years
was the contest as to the origin and maintenance of the
power of the voltaic pile. Volta himself supposed it to
reside in the contact of different metals. Here was
exerted his 'electro-motive force,' which tore the
combined electricities asunder and drove them as currents
in opposite directions. To render the circulation of the
current possible, it was necessary to connect the metals by
a moist conductor; for when any two metals were
connected by a third, their relation to each other was
such that a complete neutralization of the electric
motion was the result. Volta's theory of metallic contact
was so clear, so beautiful, and apparently so complete,
that the best intellects of Europe accepted it as the
expression of natural law.*

*Volta himself knew nothing of the chemical
phenomena of the pile; but as soon as these became
known, suggestions and intimations appeared that
chemical action, and not metallic contact, might be the*

real source of voltaic electricity. This idea was expressed by Fabroni in Italy and by Wollaston in England. It was developed and maintained by those 'admirable electricians,' Becquerel, of Paris, and De la Rive, of Geneva. The contact theory, on the other hand, received its chief development and illustration in Germany. It was long the scientific creed of the great chemists and natural philosophers of that country, and to the present hour there may be some of them unable to liberate themselves from the fascination of their first-love.

After the researches which I have endeavoured to place before you, it was impossible for Faraday to avoid taking a side in this controversy. He did so in a paper 'On the Electricity of the Voltaic Pile,' received by the Royal Society, on the 7th of April, 1834. His position in the controversy might have been predicted. He saw chemical effects going hand-in-hand with electrical effects, the one being proportional to the other; and, in the paper now before us, he proved that when the former were excluded, the latter were sought for in vain. He produced a current without metallic contact; he discovered liquids which, though competent to transmit the feeblest currents — competent therefore to allow the electricity of contact to flow through them if it were able to form a current — were absolutely powerless when chemically inactive.

By 1862, the year during which Faraday gave his last Discourse, arrangements for the Friday evening events became Tyndall's responsibility. His imaginative insights and choices of topics and speakers were much appreciated. One person of especial interest that he invited to perform was the German-born Professor at the new Royal College of Chemistry in London, August Wilhelm Hofmann (1818-1892).

Originally a student of philosophy and law, in Giessen, Hofmann became an assistant to the eminent

German chemist Justus von Liebig (one of Faraday's friends), who recommended him to Prince Albert and Queen Victoria for the London post which he held for twenty years. Hofmann was a great success in everything (scientific) that he did. He published hundreds of papers in organic chemistry, concentrating especially on products obtained from coal tar and an aniline and its reactions as well as the rearrangements exhibited by organic compounds containing nitrogen. His pupils included the great William Henry Perkin (1838-1907), the chemist who prepared the first synthetic dye, mauve. (Perkin entered the Royal College of Chemistry at the age of fifteen to study under Hofmann.) Hofmann's writings were also refreshingly devoid of the petty jealousies and rancorous polemics that characterized the exchanges between many of his contemporary organic chemists. Hofmann's private life was somewhat chequered and sad. He was four times married, and only eight of his eleven children survived him.

On Friday 11 April 1862, with Faraday in attendance, Hofmann gave a stirring account of 'Mauve and Magenta', the beautiful colouring matter derived from coal. That Discourse ended on a high note.

The material which I had to condense, I might almost say to force, into the short space of an hour, has been overwhelming; and whilst explaining the formation of the various substances which I had to describe, whilst illustrating their properties by experiment, I have scarcely had time to glance at the history of our subject. This history is not without interest. You readily perceive that a branch of industry like the one I have endeavoured to sketch could not possibly have risen like Minerva from the head of Jupiter — sudden inspiration happily realized. The time, the toil, the though of a host of inquirers were necessary to accomplish so remarkable an achievement. *You cannot expect me at this late hour to examine minutely into this part of the subject, but I must not take leave of you without alluding to some facts which cannot fail to rivet the lively interest of the Members of this Institution. Let me tell you then that Mauve and Magenta are essentially Royal Institution colours; the foundation of this new industry was laid in Albemarle Street. Benzol, which I have so repeatedly mentioned, — benzol,[6] which may be looked upon as the raw material, capable, under the influence of chemical agents, of assuming such wonderful shapes, — benzol is the discovery of our great master, may I not add of our kind friend, Mr. Faraday. This volume, The Philosophical Transactions for 1825, contains the description of his experiments. In 1825, thirty-seven years ago, the laboratory of the Royal Institution witnessed the birth of this remarkable body. Yesterday, under the auspices of Mr. Anderson, I invaded the same laboratory, a diligent search was made, and in my hand I hold the trophies of our expedition, the original specimens of benzol which Mr. Faraday prepared. In thus reminding you of one of the early labours of Mr. Faraday, — which, owing to the number and vastness of his subsequent discoveries, appears almost to have escaped from his memory like a tradition of years gone by, — I have opened a glorious page in the glorious history of the Royal Institution. Benzol has furnished us Mauve and Magenta, but it has done much more than this. Ever since chemistry became endowed with this wonderful body, benzol has been the carrier of many of the leading ideas in our science. In the hands of Mitscherlich, Zinin, Gerhardt and Laurent, in the hands of Charles Mansfield — never to be forgotten by his friends — and many others, benzol has been a powerful lever for the advancement of chemical science. Benzol and its derivatives form one of the most interesting chapters in organic chemistry, the progress of which is intimately allied with the history of this compound.*

But what has the history of benzol to do with the moral of Mauve and Magenta? Well, ladies and gentlemen, ask Mr. Faraday; ask him what in 1825 was his object in examining benzol. I have perhaps no right to answer this question in Mr. Faraday's presence; but I venture to say that we owe his remarkable inquiry to the pure delight he felt in the elaboration of truth. It was in the same spirit that his successors continued the work. Patiently they elicited fact after fact; observation was recorded after observation; it was the labour of love performed for the sake of truth; ultimately, by the united efforts of so many ardent inquirers, exerted year after year in the same direction, the chemical history of benzol and its derivatives had been traced. The scientific foundation having thus been laid, the time of application had arrived, and by one bound, as it were, these substances, hitherto exclusively the property of the philosopher, appear in the market-place of life.

Need I say any more? The moral of Mauve and Magenta is transparent enough. I read it in your eyes, — we understand each other. Whenever in future one of your chemical friends, full of enthusiasm, exhibits and explains to you his newly-discovered compound, you will not cool his noble ardour by asking him that most terrible of all questions, 'What is its use? Will your compound bleach or dye? Will it shave? May it be used as a substitute for leather?' Let him quietly go on with his work. The dye, the lather, the leather will make their appearance in due time. Let him, I repeat it, perform his task. Let him indulge in the pursuit of truth,— of truth pure and simple, — of truth not for the sake of Mauve, not for the sake of Magenta — let him pursue truth for the sake of truth!

Two influential 19th century inventors whom Tyndall invited to the Royal Institution towards the end of his period at the helm were the Welshman William Henry Preece (1834-1913), the electrical engineer who made wide-ranging contributions to the development of telegraphy, telephony and radio-telegraphic communication, and Eadweard James Muybridge (1830-1904), the English-American investigator of animal locomotion by photography.

During his long career at the Post Office, Preece was responsible for many improvements in telegraphy and in improving safety on the railways by his development of new methods of signalling. But perhaps one of his most important contributions was his encouragement of the young Italian pioneer of radio, Gugliemo Marconi (1874-1937). Preece's Discourse on 'The Telephone', given on 1 February 1878 makes fascinating reading. As well as reflecting his admiration for Faraday his lecture cites the important landmarks achieved by Edison, Bell, Helmholtz and others.

> *The telephone is an instrument constructed for the transmission of sound to a distance. The art of conveying sound to a distance is as old as the ancient Sphinx. That marvellous people the Greeks practised it; and doubtless it has served to inspire awe in many a poor pagan of simple faith as he reverently kneeled before his idol of stone or of bronze. The earliest authentic account of the germ of the telephone was within the historical period of science shadowed forth by Robert Hooke in 1667, who said:—*

'Tis not impossible to hear a whisper at a furlong's distance, it having been already done; and perhaps the nature of the thing would not make it more impossible, though that furlong should be ten times multipl'd. And though some famous Authors have affirm'd it impossible to hear through the thinnest plate of Muscovy glass; yet I know a way, by which 'tis easie enough to hear one speak through a wall a yard thick. It has not yet been thoroughly examin'd how far Otacousticons may be improv'd, not what other wayes there may be of quickning our hearing, or conveying sound through other bodies than the Air: for that that is not the only medium I can assure the Reader, that I have, by the help of a distended wire, propagated the sound to a very considerable distance in an instant, or with as seemingly quick a motion as that of light, at least incomparably quicker than that which at the same time was propagated through Air; and this not only in a straight line, or direct, but in one bended in many angles.'

This fancy remained an idea until 1819, when Wheatstone produced his 'magic lyre,' which was exhibited to delighted crowds at the Adelaide Gallery, which was often used by Professor Faraday and which has frequently since been produced by Professor Tyndall at the Royal Institution. A large musical box was placed in one of the cellars of the Institutiton, and a light rod of deal rested upon it. No sound was heard in the theatre until a light tray or other sounding-box was placed by the writer on the rod, when its music pealed forth over the whole place.[7] This was the first telephone, and was the precursor of those elegant toy-telephones which are now sold in the streets for a penny.

The vibrations of the musical box, with all their complexity and beauty, are imparted to the rod of wood, and are thence given up to the sounding-box. The sounding-box impresses them upon the air, and the air conveys them to the ear, whence they are transmitted to the brain, imparting those agreeable sensations called music. Sonorous vibrations, whether the result of music, of the human voice, or of mere noise, vary in pitch, in loudness, and in clangtint. The <u>*pitch*</u> *of a note is dependent on the length of its sonorous vibration, or on the number of sound-waves which enter the ear per second; the loudness of a note depends on the amplitude of the air-wave, or on the length of swing to and fro of the particles of air in vibration; the* <u>*clangtint,*</u> *or quality of the note, depends upon the form or rate of motion of* these particles. The limits of the ear to the reception of notes are between 16 and 38,000 vibrations per second, and the limits of the human voice are between 65 and 1044 vibrations per second. The amplitude of vibrations is very small. Lord Rayleigh has shown that a motion of only 1/25 000 000th of an inch is sufficient to produce audible sounds.

Vibrations of matter are essential to the production of sound, and the presence of air is essential to convey it to our ears.

It is possible to catch up these sonorous vibrations by placing elastic matter in their path. Thus glasses can be cracked by a loud bass voice; bodies are made to rattle in a room where music is going on; and it is only necessary to sing into a piano to receive back responsive sounds. A thin copper disc held before the mouth vibrates to each sound uttered, and if a hard metallic point be adjusted near it, sounds loud enough to fill the theatre are emitted. Acting upon this, Mr. W.H. Barlow, C.E., produced before the Royal Society in 1874 his logograph, which recorded in varying lines and curves spoken language. Here is such a line, which records the pitch, loudness, and form of the sounds emitted by the lips of the speaker, and reproducing all the elements of the voice. (See figure 51.)

In tracing the emergence of the phonograph, Preece has this to say:

It has been shown that discs vibrate under the influence of sonorous vibrations, and that these vibrations can be recorded. If these records be made on some yielding inelastic mass like tin-foil, they not only become permanent records, but they can be made to cause a similar disc at any future time to repeat or reproduce similar vibrations. Mr. T.A. Edison, of New York, has succeeded on this principle in constructing a 'Phonograph,' which repeats the voice of the speaker.

He has crystallized into a fact the ideal of the poet who longed 'for the sound of a voice that is still.'

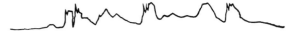

The minstrel boy to the war is gone,

In the ranks of death you'll find him ;

His father's sword he has girded on,

And his wild harp slung behind him.

Figure 51 The pitch, loudness and clangtint (quality of the note) in the transmission of sound were illustrated through this logograph shown by William Preece in his 1878 Discourse on 'The Telephone'.

Then, in describing Bell's seminal contributions, Preece gave a beautifully penetrating account of the ridicule, resistance and obloquy heaped upon inventors and discoverers:

The Phonograph is the outcome of the articulating telephone. Though several have added their share in perfecting the 'far-speaker,' there is no name in connection with it that will shine with greater brilliancy than that of Alexander Graham Bell. His father's occupation of a vocal physiologist led him to study the vocal organs and the production of sound. Helmholtz's researches led him to investigate electricity and its application to telegraphy. The desire to increase the capacity of wires for the conveyance of messages led him to devise systems of multiplex telegraphy, and this by steady and sensible degrees led him to articulate telephony. We have here a notable example of the modern method of research where imagination suggests experiment, and experiment by evolution produces growth and perfection. Things that are new are not always accepted as true. The accounts of the telephone were received in this country with great scepticism. Many even now doubt its truth until they actually test its reality. When once, however, a new thing is shown to be true, a host of detractors delight in proving that it is not new. The inventor has to pass through the ordeal of abuse. He is shown to be a plagiarist or a purloiner, or sometimes worse. Others are instanced as having done the same things years ago, though perhaps their own existence apart from their ideas have never before been heard of. Professor Bell will have to go through all this; nevertheless the telephone will always be associated with his name, and it will remain one of the marvels of this marvellous age, while its chief marvel will be its beautiful and exquisite simplicity.

Eadweard Muybridge, born Edward James Muggeridge, in Kingston-on-Thames, near London in 1830, changed his name because he believed it to be the original Anglo-Saxon form. He emigrated to California in 1852, became a professional photographer about 1866. He made quite a reputation through his impressive views of Yosemite; and soon became the chief photographer to the U.S. Government. Leland Stanford, ex-Governor of California, following arguments about the gait of a horse, commissioned

Muybridge in 1872 to make a photographic investigation at his stud farm. Using a battery of some twenty-four small cameras with electromagnetic shutters set off by clockwork or by threads stretched across the track which were broken by the running horse, Muybridge exposed the fallacy of the 'rocking-horse' attitude depicted by artists. The zoopraxiscope was the name given by Muybridge to he lantern that he developed. This projected images in rapid succession onto a screen from photographs printed on a rotating glass disc. This produced the illusion of moving pictures, which is why Muybridge is often called the stepfather of modern motion pictures.

A special Monday evening Discourse was arranged at the Royal Institution on 13 March 1882 when, in the presence of HRH The Prince of Wales (chairman for the event), Princess Alexandra, Gladstone, Huxley, Tennyson and Tyndall, Eadweard Muybridge spoke on 'The Attitudes of Animals in Motion, Illustrated with the Zoopraxiscope.' Members and their guests saw that night images such as those shown in figure 52. For historical, cultural and scientific reasons excerpts of his Discourse are reproduced below.

The problem of animal mechanism has engaged the attention of mankind during the entire period of the world's history.

Job describes the action of the horse; Homer, that of the ox; it engaged the profound attention of Aristotle, and Borelli devoted a lifetime to its attempted solution. In every age, and in every country, philosophers have found it a subject of exhaustless research. Marey, the eminent French savant of our own day, dissatisfied with the investigations of his predecessors, and with the object of obtaining more accurate information than their works afforded him, employed a system of flexible tubes, connected at one end with elastic air-chambers, which

SOME CONSECUTIVE PHASES OF THE GALLOP.

Figure 52 An example — in this case some consecutive phases of the gallop of a horse — of some of the results shown by Muybridge at the Royal Institution in 1882.

were attached to the shoes of a horse; and at the other end with some mechanism, held in the hand of the animal's rider. The alternate compression and expansion of the air in the chambers caused pencils to record upon a revolving cylinder the successive or simultaneous action of each foot, as it correspondingly rested upon or was raised from the ground. By this original and ingenious method, much interesting and valuable information was obtained, and new light thrown upon movements until then but imperfectly understood.

While the philosopher was exhausting his endeavours to expound the laws that control, and the elements that effect the movements associated with animal life, the artist, with a few exceptions, seems to have been content with the observations of his earliest predecessors in design, and to have accepted as authentic without further inquiry, the pictorial and sculptural

169

representations of moving animals bequeathed from the remote ages of tradition.

When the body of an animal is being carried forward with uniform motion, the limbs in their relations to it have alternately a progressive and a retrogressive action, their various portions accelerating in comparative speed and repose as they extend downwards to the feet, which are subjected to successive changes from a condition of absolute rest, to a varying increased velocity in comparison with that of the body.

The action of no single limb can be availed of for artistic purposes without a knowledge of the synchronous action of the other limbs; and to the extreme difficulty, almost impossibility, of the mind being capable of appreciating the simultaneous motion of the four limbs of an animal, even in the slower movements, may be attributed the innumerable errors into which investigators by observation have been betrayed. When these synchronous movements and the successive attitudes they occasion are understood, we at once see the simplicity of animal locomotion, in all its various types and alternations. The walk of a quadruped being its slowest progressive movement would seem to be a very simple action, easy of observation and presenting but little difficulty for analysis, yet it has occasioned interminable controversies among the closest and most experienced observers.

When, during a gallop, the fore and hind legs are severally and consecutively thrust forwards and backwards to their fullest extent, their comparative inaction may create in the mind of the careless observer an impression of indistinct outlines; these successive appearances were probably combined by the earliest sculptors and painters, and with grotesque exaggeration adopted as the solitary position to illustrate great speed. Or, as is very likely, excessive projection of limb was intended to symbolise speed, just as excess in size was an indication of rank. This opinion is to some extent *corroborated by the productions of the Grecian artists in their best period, when their heroes are represented of the same size as other men, and their horses in attitudes more nearly resembling those possible for them to assume. The remarkable conventional attitude of the Egyptians, however, has, with few modifications, been used by artists of nearly every age to represent the action of galloping, and prevails without recognised correction in all civilised countries at the present day.*

The ambition and perhaps also the province of art in its most exalted senses, is to be a delineator of impressions, a creator of effects, rather than a recorder of facts. Whether in the illustrations of the attitudes of animals in motion the artist is justified in sacrificing truth, for an impression so vague as to be dispelled by the first studied observation, is a question perhaps as much a subject of controversy now as it was in the time of Lysippus, who ridiculed other sculptors for making men as they existed in nature; boasting that he himself made them as they ought to be.

A few eminent artists, notable among whom is Meissonier, have endeavoured in depicting the slower movements of animals to invoke the aid of truth instead of imagination to direct their pencil, but with little encouragement from their critics; until recently, however, artists and critics alike have necessarily had to depend upon their observation alone to justify their conceptions or to support their theories.

Photography, at first regarded as a curiosity of science, was soon recognised as a most important factor in the search for truth, and its more popular use is now entirely subordinated by its value to the astronomer, the anatomist, the pathologist, and other investigators of the complex problems of nature. The artist, however, still hesitates to avail himself of the resources of what may be at least acknowledged as a handmaiden of art, if not admitted to its most exalted ranks.

Having devoted much attention in California to experiments in instantaneous photography, I, in 1872, at the suggestion of the editor of a San Francisco newspaper, obtained a few photographic impressions of a horse during a fast trot.

At this time much controversy prevailed among experienced horsemen as to whether all the feet of a horse while trotting were entirely clear of the ground at the same instant of time. A few experiments made in that year proved a fact which should have been self-evident.

Being much interested with the experiments of Professor Marey, in 1877 I invented a method for the employment of a number of photographic cameras, arranged in a line parallel to a track over which the animal would be caused to move, with the object of obtaining, at regulated intervals of time or distance, several consecutive impressions of him during a single complete stride as he passed along in front of the cameras, and so of more completely investigating the successive attitudes of animals while in motion than could be accomplished by the system of M. Marey.

I explained the plan of my intended experiments to a wealthy resident of San Francisco — Mr. Stanford — who liberally agreed to place the resources of his stock-breeding farm at my disposal, and to reimburse the expenses of my investigations, upon condition of my supplying him, for his private use, with a few copies of the contemplated results.

Some of the illustrations shown that night are given in figures 53 and 54. And Muybridge's closing paragraph had a prophetic ring:

The employment of automatic apparatus for the purpose of obtaining a regulated succession of photographic exposures is too recent for its value to be properly understood, or to be generally used for scientific experiment; at a future time, the pathologist, the anatomist, and other explorers for hidden truths will find it indispensable for their complex investigations.

Figure 53 Muybridge used wires that were actuated by the galloping horse to time the exposure necessary to photograph the moving animal.

Figure 54 Another of the figures (like figure 53) shown by Muybridge during his Discourse to illustrate a similar point.

It is interesting to reflect that another Californian, the Egyptian-born Ahmed Zewail (Linus Pauling Professor at the Californian Institute of Technology) began his Friday evening Discourse in March 1991 with reference to Muybridge's work. In 1882, the exposures used in the zoopraxiscope were about a thousandth of a second. Using the fastest currently available pulsed lasers, Zewail's studies of chemical reactions utilize 'photographs' taken in a millionth of (an American) billionth (10^{-15}) of a second, that is on the femtosecond time scale. There are as many femtoseconds in a second, as there are seconds in 32 million years.

Sir James Dewar and Lord Rayleigh

It has been rightly said[8] that 'the Royal Institution has taken its character from its Professors; to them it owes its reputation.' James Dewar (figure 55) first took up the Fullerian Professorship of Chemistry in 1887, twelve years after he had been appointed Jacksonian Professor of Natural Philosophy at the University of Cambridge: he held both posts for over fifty years! The son of a Scottish Lowland publican, great chemist, brilliant experimentalist, accomplished musician, keen astronomer, lover of poetry and an irrepressible inventor and gadgeteer, Dewar was also a difficult, fractious individual. He and Tyndall overlapped at the Institution for ten years. At first, they got on well together, and Dewar would dine with the Tyndall's when he came to lecture. Gradually, however, their relationship became strained. It did not help that whilst Tyndall was moody with insomniac tendencies, Dewar, by nature somewhat cantankarous, suffered perpetually from indigestion. Trouble developed between them and the ageing Tyndall could not accept the increasingly assertive and vigorous style adopted by Dewar. When Tyndall was not allowed (by the Managers) to give the Christmas Lectures in 1887-8, and they were offered instead to Dewar, it was the last straw. Tyndall threw in his hand and resigned. Dewar was made Superintendent of the House, and rather hustled the Tyndalls out of the Director's flat.

Figure 55   Sir James Dewar; Fullerian Professor of Chemistry at the Royal Institution from 1877 to 1923.

Notwithstanding his many personal faults Dewar had a touch of genius as a scientist, and in his choice of interesting Discourse speakers he was excellent. Poets, musicians, actors and artists as well as the leading scientists of the day appeared in his programme of events. Sir Henry Irving (1838-1905), the great actor, who lived for nearly thirty years in a corner flat situated less than 100 yards from the Institution's lecture theatre, gave a special afternoon Discourse on 1 February 1895 on 'Acting: An Art', attracting an audience of 1050. There were numerous other eminent performers, including the Director of the Royal College of Music, Sir Hubert Parry, who composed the tune to which William Blake's 'Jerusalem' is sung on the last night of the Royal Albert Hall Promenade concerts.

Dewar was friendly with Pierre and Marie Curie who frequently visited him at the Institution. The former gave a Discourse on Radium (in French) in 1903. He also persuaded other leading French scientists (including Moissan and Perrin) to give Discourses. The eighty year old Alfred Russel Wallace spoke on Darwinism and his own independent seminal contributions to evolution in 1909; the young H G Wells expatiated on 'The Discovery of the Future' in 1902 (see page 207); and the great American astronomers George Ellery Hale (1868-1938) and Percival Lowell (1855-1916) spoke in successive years. Hale dealt with 'Solar Vortices' and Lowell with his observatory's 'Photographs of the Planets.' Dewar had an emotional affinity with Lowell, whose early life was devoted to literature and travel. It was not until his mid-thirties, inspired by what he had heard of the 'discovery' of canals on Mars that Lowell turned to astronomy and set out to build his famous private observatory at Flagstaff, Arizona. (Lowell predicted the existence of the planet Pluto and initiated the search that ended in its discovery fourteen years after his death. Only after the dawn of the space age in the second half of the 20th century were Lowell's views about the canals on Mars finally laid to rest.)

One particular (extra) Discourse organized by Dewar took the intellectual and scientific life of London by storm. This was the two-hour special presentation, inserted in the Institution's programme at the last minute, by the remarkable Croatian electrical engineer Nikola Tesla (1856-1943) who revolutionized power transmission in industry and who, for most of his life, was obsessed with the idea of transmitting energy without wires.[9] There are probably more legends about Tesla than about any other experimenter in electrical science. Tesla spoke, in February 1892 on 'Currents of High Potential and of High Frequency.' He excited the leader writers in the serious daily press. One of them wrote:

Votaries of other sciences frequently complain at the present day that the fascinations of electrical research and the substantial rewards of applied electrical science are attracting a disproportionate amount of the intelligence of the rising generation. They say that few young men are found to devote themselves, say to astronomy, or biology, while they flock to the study of electrical engineering. We may probably take it that this is more or less the case, and the explanation is not very far to seek. Electricity has laid hold of the imagination of the age. It enters into every department of life, and always with the effect of simplifying, accelerating, and improving the arrangements to which we had been accustomed. The newest torpedo is driven, controlled, and exploded by electricity; and the electric motor is the destined solution of half our domestic worries. Alike among serious students and among the general public there is strong faith in the possibilities of the science, which offers at once one of the richest and most attractive of modern fields of research, and one of the most promising applications of science to the acquisition of wealth. Some of the sciences now complaining of neglect do not offer the same bread-and-butter attractions, but this, we believe, is a less serious drawback than their want of scientific outlook that electrical engineering possesses. In some directions we are too visibly reaching, not indeed exhaustion of what is to be known, but the limits of the means of research within our reach. We cannot hope, for example, to see anything much smaller than our present microscopes reveal, nor can we hope for much solid addition to the information given by our best telescopes. But every one feels that we are only on the threshold of knowledge about electricity, whether abstract or applied, and that feeling has determined a flow of young men to the workshops of electrical engineers which cannot be financially justified save by a very rapid development of the comparatively youthful science.

If anything were needed to stimulate enthusiasm or strengthen faith, it was surely supplied last night by the very remarkable lecture with which Mr. Tesla for two hours held a professional audience entranced at the Royal Institution. His beautiful experiments not only open up a new and most promising field of investigation, but also suggest more or less distinctly a revision of many general physical conceptions and a stimulating expansion of our speculative ideas. Mr. Tesla is working on the borderland where light, heat, electricity, chemical affinity, and forms of energy which we cannot confidently identify with any of these, meet and blend. Watching some of his striking experiments, one feels that old lines of demarcation are fading away, and that some new and fruitful generalization cannot be very far off with which we may start upon new voyages of discovery. To take the most obvious and elementary reflection, the spectator involuntarily asks himself what precisely is meant by electrics and di-electrics, by conducting and insulating bodies. Mr. Tesla establishes an arc between two poles, then passes between them a plate of the best di-electric we know of, and the result is not to check, far less to interrupt, the discharge, but positively to facilitate it. In other experiments of the same kind he shows his high potential currents absolutely disregarding all the devices by which ordinary currents are held in check. It seems to follow that there is no di-electric, no thickness of ebonite, and no space of air that cannot be pierced or bridged by a current of appropriate intensity. It may not seem a remarkable discovery when once it is stated, but nevertheless it is one of the things that people do not realize, and it breaks down quite a number of conceptions which, in the long run, rest upon the tacit assumption that there are hard and fast division lines. Upon this follows the remarkable discovery that as electricity increases in physical power it loses its effect upon the human frame. The lecturer stands in an electrostatic field capable of setting a lamp aglow without wires and feels nothing. He puts one hand upon a terminal from which a brush of violet discharge is crackling and spluttering, in the other he holds a lamp or a vacuum tube, thus making himself the channel for a current at something like 50,000 volts. The vacuum tube glows like the gates of sunrise, but the lecturer feels nothing, though a current of one five-hundredth of the intensity might easily terminate his career. As the sun pours forth rays which the human eye cannot see, so it would seem that the flux of energy in which we live and move finds no response in the human nerve until it is slowed down to the pace of a common dynamo.

For the practical man this is really a most comfortable discovery. The practical man naturally wants to know what he is to get out of all this, and whether electric lighting is forthwith to become cheaper than gas. It is, therefore, no small thing from his point of view that his drawing-room may be converted into a electrostatic field so that lamps and vacuum tubes anywhere in it shall emit any desired radiance, while he moves in perfect safety amid the *storm of molecular energy which defies the insulating powers of glass. Mr. Tesla dropped various practical hints in the course of his lecture, but he did not precisely respond to the desire of the practical man for cheaper lights.*

Ten years after the arrival of Dewar at the Royal Institution, another even more eminent Cambridge scientist migrated there, Lord Rayleigh (John William Strutt, 1842-1919), who was appointed successor to Tyndall as Professor of Natural Philosophy in 1887.

Upon Clerk Maxwell's death in 1879, Rayleigh was elected his successor as Cavendish Professor of Physics in the University of Cambridge. He resigned from that post in 1884 to attend to his estate at Terling in Essex, where he set up a well-equipped private laboratory (figure 56). For the remainder of his life he pursued research work there and at the Royal Institution where he remained as Professor of Natural Philosophy until 1905. He gave numerous Discourses, and conducted research on an extraordinarily wide range of topics. He explained the blue of the sky, he pioneered studies in sound and surface acoustical waves, optics, fluid mechanics and colloid science. Above all, however, he discovered argon, for which work he was awarded — with William Ramsay, James Dewar's adversary, of University College, London — the 1904 Nobel Prize. Rayleigh, the first of fifteen Professors of the Royal Institution to become Nobel laureates, would frequently bring his brother-in-law, The Rt Hon A J Balfour, Prime Minister and later Foreign Secretary (and author of the Balfour Declaration that led to the creation of the State of Israel) to Friday Evening Discourses. They are each to be seen along with other luminaries, in the front row of the audience listening to James Dewar presenting his famous 1904 Discourse on 'Liquid Hydrogen' (figure 57).[10]

Figure 56 Lord Rayleigh, Professor of Natural Philosophy 1887-1905, Honorary Professor 1905-1919, at his bench.

Both Dewar and Rayleigh maintained the tradition of bringing to the attention of the Royal Institution the most recent advances in science and contiguous areas of intellectual activity. This had been very much the ethos that had motivated Faraday. When J J Thomson

(1856-1940), the Cavendish Professor at Cambridge (Rayleigh's successor in that post), was invited by Dewar to present a discourse in 1897, his announcement that night of the discovery of the electron was greeted with some scepticism.

Figure 57 James Dewar's Discourse on liquid hydrogen in 1904. Two future Nobel prizewinners (Rayleigh and Marconi), a Prime Minister (Balfour) and four Presidents of the Royal Society, including Kelvin, Stokes and Crookes, as well as Ludwig Mond and George Malthey, captains of industry, were in the audience.

In his *Recollections and Reflections,* published in 1936, Thomson recalled:

At first there were very few who believed in the existence of these bodies smaller than atoms. I was even told long afterwards by a distinguished physicist who had been present at my lecture at the Royal Institution that he thought I had been 'pulling their legs'.

Lord Rayleigh described at Discourses and afternoon or lunchtime lectures his Nobel-prize-winning discovery of the noble gas, argon, his pioneering investigations of sound, radiation, fluid flow and numerous other topics. It was Rayleigh who first explained why Faraday's colloids (pages 78-80) were coloured, and he did so quantitatively, in terms of the dielectric properties that Faraday himself had discovered.[11] Dewar described his own ingenious ways of attaining low temperature and of liquefying the permanent gases. In 1926, Logie Baird (1888-1946) gave the first demonstration of television; and in 1929 Leonard Woolley first disclosed his archaeological findings in Ur of the Chaldees. In 1931 Maynard Keynes described 'The Internal Mechanics of the Trade Slump'. In 1932 Lord Rutherford announced the discovery of the neutron by his associate James Chadwick at the Cavendish Laboratory. Rutherford's enthusiasm was always infectious; everyone loved him whether they understood his Discourse or not, as he boomed and coughed, and the lock of white hair flopped on his forehead[12]. In the early 1930s W H Bragg (then Director of the Institution) and his former colleague at the Davy Faraday laboratory, W T Astbury,[13] announced their major discovery that hair, wool and other natural fibres were crystalline, a breakthrough which marked the beginning of structural molecular biology. This was as momentous in its own right as the demonstration by Wohler in the 1820s that some 'organic' molecules (notably urea) may be produced from entirely inorganic precursors. Thereafter a sequence of major discoveries have been announced at Friday Evening Discourses, including new insights into silicate minerals and alloys (by Sir Lawrence Bragg); the nature of the liquid state (by J D

(a)

Friday, November 1

M. F. PERUTZ, C.B.E., Ph.D., F.R.S., and
J. C. KENDREW, C.B.E., Sc.D., F.R.S.

*Chairman and Deputy Chairman of the Medical Research Council
Laboratory of Molecular Biology, Cambridge, and Readers in the
Davy Faraday Research Laboratory*

STRUCTURE OF PROTEINS

Proteins are vital components of every living cell and one of the
fundamental problems of biochemistry is to understand their
modes of functioning. Protein molecules are so complex, containing thousands of atoms, that the classical methods of chemistry
are inadequate. It has been necessary to invoke a physical technique—X-ray crystallography—and even with this, protein structures were only solved after many years of study and then by
stages. The solutions of the structures of myoglobin and haemoglobin, which are to be described in the Discourse, were the first
successful projects in this field.

Myoglobin, like haemoglobin, combines reversibly with oxygen
and is a relatively small protein (mol. wt. 17,000) found in muscle
cells. Haemoglobin is four times as large, and apart from general
principles of protein architecture, its structure is of great physiological interest, in view of its function as a carrier of oxygen and
carbon dioxide in the blood.

*(Members will recall that Dr. Perutz and Dr. Kendrew, who are
Readers in the Davy Faraday Laboratory, shared the Nobel Prize
for Chemistry last year for their work on protein research, and that
there has been very close collaboration between their laboratory and
ours).*

Max F. Perutz

Chairman of the Medical Research Council Laboratory of Molecular Biology since 1962
and Director of the Unit bearing the same name since 1947. Educated at the Universities
of Vienna and Cambridge. I.C.I. Research Fellow at the University of Cambridge 1945-
47. Lecturer in Biophysics at the University of Cambridge 1953-57. Reader, Davy
Faraday Research Laboratory, since 1955. Published papers on the crystal structure of
proteins in various journals. F.R.S. 1954. Nobel Prize in Chemistry 1962. C.B.E. 1963.

John C. Kendrew

Deputy Chairman of the Medical Research Council Laboratory of Molecular Biology
since 1962 and Deputy Director of the Unit bearing the same name since 1947. Fellow of
Peterhouse, Cambridge. Reader, Davy Faraday Research Laboratory, since 1954.
Educated at Clifton College and Trinity College, Cambridge. Member of the

(b)

(c)

Figure 58 On the facing page can be seen (a) the announcement of the historic joint Discourse given by two Nobel-prize-winning members of staff of the Davy Faraday Research Laboratory, and (b) Dr John Kendrew in action that night, 1 November 1963 on 'The Structure of Proteins'. (c), above, shows the other member of the pair, Max Perutz, as drawn by his mentor, Sir Lawrence Bragg.

Bernal who, in his younger days working along W H Bragg at the Institution, solved the structure of the lubricating mineral used in pencils, graphite). The Nobel-prize-winning breakthroughs concerning the structure of haemoglobin and myoglobin by Max Perutz and John Kendrew, resulted in one of the rare Discourses (in 1963) when there were two speakers (see figure 58). David C Phillips , a key member of staff at the Davy Faraday Laboratory in W L Bragg's day as Director, achieved a world first in 1965 by solving the

185

structure of an enzyme (lysozyme). His Discourse in November 1965 showed for the first time how X-ray crystallography, which was largely founded at the Royal Institution, could explain the mechanism of the mode of action and function of an enzyme.

Included among the ranks of distinguished European scientists who lectured at the Royal Institution in the 19th century were many colourful characters such as the Russian educational reformer and economist D I Mendeleeff (figure 59), better known as the formulator of the Periodic Law of the elements; the great organic chemist J B Dumas (a lifelong friend of Faraday's) who, after the revolutionary changes in France in 1848, became that country's Minister of Agriculture, and later Minister of Education; and S Cannizzaro the Italian chemist and one-time member

(a)

(b)

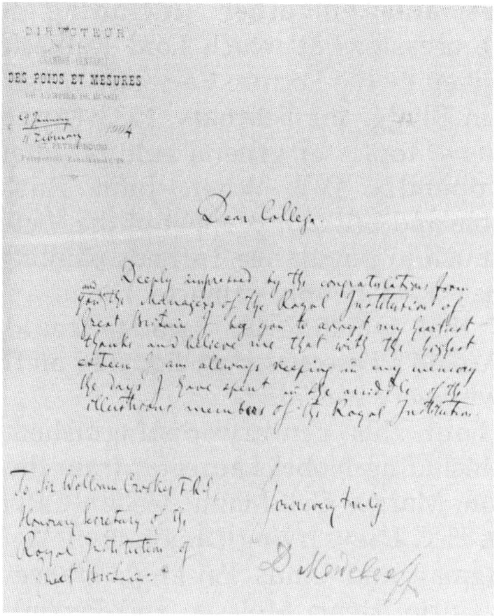

DES POIDS ET MESURES

Dear Collega.

Deeply impressed by the congratulations from the Managers of the Royal Institution of Great Britain I beg you to accept my heartiest thanks and believe me that with the highest esteem I am always keeping in my memory the days I have spent in the midst of the illustrious members of the Royal Institution.

To Sir William Crookes F.R.S. Honorary Secretary of the Royal Institution of Great Britain.

Yours very truly

D Mendeleeff

Figure 59 On facing page, (a) shows Dmitri Ivanovitch Mendeleeff (1834-1907), the fourteenth and last son of the Director of the Gymnasium at Tobulsk. A prolific author, of more than 300 original papers in art, education and economics, his greatest triumph was to bring forth the Periodic Table of the elements. He gave a Discourse at the Royal Institution in 1889, and thereafter corresponded regularly with members of the Royal Institution. (b), above, is a letter of thanks from Mendeleeff to Sir William Crookes (from the archives of the Royal Institution).

of Garibaldi's military forces, distinguished for resurrecting Avogadro's hypothesis. An indication of international cooperation involving scientists at the Royal Institution is conveyed by the list of those who were elected Honorary Members of the Institution in 1891.[14] Insight into the range of topics covered both at

Discourses and on other (including Saturday afternoon), occasions at which Lord Rutherford often spoke, is given by the Lecture Calendar, constructed by Sir William Bragg, for February 1933 (Appendix V). Then, as now, topics of general cultural interest were aired (Appendix IV). When John Ruskin, the indefatigable and brilliant prophet of the Victorian age who did much to popularize Turner's paintings, spoke on 'Verona', 1144 people came to listen, a figure in excess of that which even Faraday could attract. Mathew Arnold's Discourses on Equality and Emerson were likewise very well attended.

Throughout this century distinguished foreign scientists including Nobel Laureates from the USA (A H Compton, Murray Gell-Mann, Melvin Calvin, Roald Hoffmann, H C Urey, Irving Langmuir, T W Richards, P W Bridgman and Linus Pauling); France (Gabriel Lippman, Curie, Henri Moissan and Perrin); Austria (Konrad Lorenz, Nikolaas Tinbergen and Erwin Schroedinger); Netherlands (Lorentz and Zeeman); Sweden (Arrhenius); Germany (Ostwald and Eigen); Italy (Emilio Segre); Spain (S Ochoa); and other countries have given Discourses at the Royal Institution. The lecture theatre also carries echoes of the presence of other great men and women of the past that inspire each new generation: Walford Davies and Parry, Gilbert Murray, G M Trevelyan, Sir James Frazer, Jacquetta Hawkes, Dorothy Sayers, Kathleen Kenyon, Anthony Blunt, Margaret Mead, Basil Spence, Gerald Kelly, Peter Scott, Osbert Lancaster, Gerald Moore and Hugh Casson.

Notes

1. John Hanning Speke (1827-1864), the other great explorer associated with the discovery of the source of the Nile, also gave a special Tuesday evening Discourse on a similar topic in 1863. Speke died in a tragic shooting accident on the day he was scheduled to debate his claims publicly with yet another great explorer, Sir Richard Burton.

2. I M McCabe, *Proc. Royal Inst. of G.B.*, **61**, 283 (1989).

3. Described by Sir Antony Jay, broadcaster and writer, and author of *Yes Minister* and *Yes Prime Minister*, as 'the best studio in the world'. (Sir Antony gave a Discourse on 'Understanding Laughter' in 1990, see *Proc. Royal Inst. of G.B.*, **62**, 99, (1990).)

4. The 'Discourse' was subsequently shown on New Zealand national television and in the schools of the United Kingdom and other countries.

5. A delightful account of the life of Charles Piazzi Smyth entitled 'The Peripatetic Astronomer' by H A Brück and M T Brück has recently appeared (Adam Hilger, Bristol, 1988).

6. Benzol is now called benzene.

7. This demonstration was performed during Professor Charles Taylor's Christmas Lectures (1988-89). See also Charles Taylor, *The Art and*

Science of Lecture Demonstration, Adam Hilger, Bristol, (1988), 35.

8. Gwendy Caroe, *The Royal Institution: An Informal History,* John Murray (1985), 86.

9. A claim was made by Tesla and by others who allegedly witnessed the event that he was able to light some 200 50-watt lamps at over twenty miles from a generator by using the earth alone as a transmitting medium.

10. Of the eminent persons in the front row, Lodge, Rayleigh, Balfour and Crookes were all deeply interested in physical phenomena.

11. See M Kerker, *Proc. of the Royal Inst. of G.B.,* 61 (1989), 229.

12. See Mrs Alban Caroe, *Proc. of the Royal Inst. G.B.* (1966), 398.

13. For this work, Astbury, later Professor of Biophysics in the University of Leeds, earned the Actonian Prize of the Royal Institution. This prize has been awarded every seventh year since 1844. Previous recipients include Madame Curie for her studies of radioactive substances; Sir Charles Sherrington for the integrative action of the nervous system, and Sir Alexander Fleming for the discovery and development of penicillin.

14. The total number of Honorary Members is restricted to 25, including the Monarch, the Regent and President of the Royal Society. At present

there are twelve Nobel Laureates, one recipient of the Japan prize and two of the Wolf prize. Some of those elected in 1891 were Louis Pasteur, Berthelot, Bunsen, Helmholtz, Dana, Willard Gibbs, S Newcomb, Cannizzaro, Mendeleeff, Van der Waals and Hofmann. In his letter of thanks to the Secretary of the Royal Institution, Hofmann wrote:

In Faraday, I admired the incomparable experimental thinker; I loved the noble-minded, kind-hearted man. During the twenty years I have had the good fortune of living in dear old England I did not miss a single one of his lectures. It would be difficult to say how deeply I am indebted to these and to a great many other lectures I attended at the Royal Institution, and often in my own courses when showing a highly instructive experiment, I delight in telling my students where and by whom performed I saw it for the first time.

Chapter 8

The Popularization of Science

Faraday once wrote that 'a truly popular lecture cannot teach, and a lecture that truly teaches cannot be popular.' Irrespective of the veracity or otherwise of the statement, the Royal Institution continues in its endeavours to popularize, just as it did in 1826 when the practice in its present form first began. Faraday expressed the view that evening lectures should amuse and entertain as well as educate, edify and, above all, inspire. This is still the principle that governs the Royal Institution's numerous educational activities. Because of initiatives taken under the aegis of Sir Lawrence Bragg (1890-1971), Sir George (later Lord) Porter (1920-) and their successor, these activities now extend to mathematics and technology masterclasses for schoolboys and girls (given also in twenty-five other centres throughout the UK as well as at the Institution itself), to school curriculum enhancement projects, and to the making of special video films (on such topics as 'Geometry and Perspective' and 'Colour'). For young people, however, the highlight every year is the series of Christmas Lectures which, since 1966, have been televised nationwide by the BBC, and are now repeated in abbreviated form to the schoolchildren of Japan during the succeeding summer. Video copies of the Christmas Lectures are now marketed worldwide.[1]

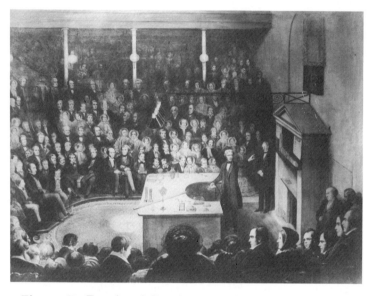

Figure 60 Faraday delivering one of his 1855 Christmas Lectures.

The most famous series of Christmas Lectures are still those given by Michael Faraday on *The Chemical History of a Candle*. (Japanese translations of this book have run to over seventy editions.) In his opening remarks in the first of his six lectures, Faraday spoke to the multitude of children assembled in the theatre of the Royal Institution in December, 1860 thus:

I have taken this subject on a former occasion;[2] and were it left to my own will, I should prefer to repeat it almost every year — so abundant is the interest that attaches itself to the subject, so wonderful are the varieties of outlet which it offers into the various departments of philosophy. There is not a law under which any part of this universe is governed which does not come into play, and is touched upon in the phenomena. There is no better, there is no more open door by which you can enter into the study of natural philosophy, than by considering the physical phenomena of a candle. I trust, therefore, I shall not disappoint you in choosing this for my subject rather than any newer topic, which could not be better, were it even so good.

And before proceeding, let me say this also — that though our subject be so great, and our intention that of treating it honestly, seriously, and philosophically, yet I mean to pass away from all those who are seniors amongst us. I claim the privilege of speaking to juveniles as a juvenile myself. I have done so on former occasions — and, if you please, I shall do so again. And though I stand here with the knowledge of having the words I utter given to the world, yet that shall not deter me from speaking in the same familiar way to those whom I esteem nearest to me on this occasion.

And now, my boys and girls, I must first tell you of what candles are made. Some are great curiosities. I have here some bits of timber, branches of trees particularly famous for their burning. And here you see a piece of that very curious substance taken out of some of the bogs in Ireland, called candle-wood, — a hard, strong, excellent wood, evidently fitted for good work as a resister of force, and yet withal burning so well that where it is found they make splinters of it, and torches, since it burns like a candle, and gives a very good light indeed. And in this wood we have one of the most beautiful illustrations of the general nature of a candle that I can possibly give. The fuel provided, the means of bringing that fuel to the place of chemical action, the regular and gradual supply of air to that place of action

— heat and light — all produced by a little piece of wood of this kind, forming, in fact, a natural candle.

These words, and those that follow them, along with the elegant sequence of simple experiments described by Faraday in the written version of his Christmas Lectures have made *The Chemical History of a Candle* a classic in the annals of science. The Preface, composed with such felicity for a later edition by William Crookes (1832-1919), a prominent member and active participant in the affairs of the Institution (and President of the Royal Society 1913-15) (figure 61), adds to the charm.

From the primitive pine-torch to the paraffin candle, how wide an interval! between them how vast a contrast! The means adopted by man to illuminate his home at night, stamp at once his position in the scale of civilisation. The fluid bitumen of the far East, blazing in rude vessels of baked earth; the Etruscan lamp, exquisite in form, yet ill adapted to its office; the whale, seal, or bear fat, filling the hut of the Esquimaux or Lap with odour rather than light; the huge wax candle on the glittering altar, the range of gas lamps in our streets, — all have their stories to tell. All, if they could speak (and, after their own manner, they can), might warm our hearts in telling, how they have ministered to man's comfort, love of home, toil, and devotion.

Surely, among the millions of fire-worshippers and fire-users who have passed away in earlier ages, some have pondered over the mystery of fire; perhaps some clear minds have guessed shrewdly near the truth. Think of the time man has lived in hopeless ignorance: think that only during a period which might be spanned by the life of one man, has the truth been known.

Atom by atom, link by link, has the reasoning chain been forged. Some links, too quickly and too slightly made, have given way, and been replaced by better work; but now the great phenomena are known — the outline is correctly and firmly drawn — cunning artists are filling in the rest, and the child who masters these Lectures knows more of fire than Aristotle did.

The candle itself is now made to light up the dark places of nature; the blowpipe and the prism are adding to our knowledge of the earth's crust; but the torch must come first.

Among the readers of this book some few may devote themselves to increasing the stores of knowledge: the Lamp of Science must burn. 'Alere flammam.'

Figure 61 Sir William Crookes, inventor of the radiometer and of the evacuated tubes named after him, discoverer of the element thallium. At various times he was President of the Chemical Society, the Institution of Electrical Engineers, the Society of the Chemical Industry, the British Association and the Royal Society.

Figure 62 Portrait of Faraday, aged about sixty, by McGuire.

When Sir James Jeans (1877-1946), the distinguished astronomer, and an expert musician, gave the Christmas Lectures in 1933-4 he made them the basis of his influential book *Through Space and Time* which subsequently inspired generations of young scientists. The preface of that book encapsulates the essence of the Christmas series:

Every year for more than a century, the Royal Institution has invited some man of science to deliver a course of lectures at Christmastide in a style 'adapted to a juvenile auditory'. In practice, this rather quaint phrase means that the lecturer will be confronted with an eager and critical audience, ranging in respect of age from under

197

eight to over eighty, and in respect of scientific knowledge from the aforesaid child under eight to staid professors of science and venerable Fellows of the Royal Society, each of whom will expect the lecturer to say something that will interest him.

The present book contains the substance of what I said when I was honoured with an invitation of this kind for the Christmas season 1933-4, fortified in places with what I said on other slightly more serious occasions, both at the Royal Institution and elsewhere.

His opening paragraphs conjured up the magic of astronomy which is the single most popular subject chosen so far in the one-hundred and sixty two series presented to date:

These are restless days in which everyone travels who can. The more fortunate of us may have travelled outside Europe to other continents — perhaps even round the world — and seen strange sights and scenery on our travels. And now we are starting out to take the longest journey in the whole universe. We shall travel — or pretend to travel — so far through space that our earth will look less than the tiniest of motes in a sunbeam, and so far through time that the whole of human history will shrink to a tick of the clock, and a man's whole life to something less than the twinkling of an eye.

As we travel through space, we shall try to draw a picture of the universe as it now is — vast spaces of unthinkable extent and terrifying desolation, redeemed from utter emptiness only at rare intervals by small particles of cold lifeless matter, and at still rarer intervals by those vivid balls of flaming gas we call stars. Most of these stars are solitary wanderers through space, although here and there we may perhaps find a star giving warmth and light to a family of encircling planets. Yet few of these are at all likely to resemble our own earth; the majority will be so different that we shall hardly be able to describe their scenery, or imagine their physical condition.

The subjects touched upon in the Christmas Lectures have ranged from music to molecules, from seeing the very small (by electron microscopy) to exploring the very large (by astronomy), from the language of animals — brilliantly expounded by David Attenborough in 1973-74 before he became a world renowned television star — to exciting adventures into both the large and small, under the laws of Gulliver (figures 63-65). Only once to date has mathematics figured centrally in the series. So successful was E C Zeeman's (Professor of Mathematics at the Royal Institution) treatment of his subject in December-January 1979-80 (**Mathematics into Pictures**) that, by popular demand, they led to the formation of regular Royal Institution Masterclasses in this subject throughout the United Kingdom. Zeeman introduced **The Nature of Mathematics and the Mathematics of Nature** in the following terms:

This is the first time in 149 years that the Christmas Lectures have been devoted to mathematics. Maybe this is because it is a paradoxical subject: we are never quite sure whether it is an art or a science, whether we invent it or discover it, whether it is a man-made toy or a truth so universal that it is independent of the universe. It is one of the oldest and most splendid endeavours of mankind. Some people love it and others hate it. It can be very pure with its patterns of growth determined by its own intrinsic criteria of beauty, or it can be very applied with its patterns of growth determined by its usefulness to science. So we shall start with a pure point of view and spend the first three lectures looking at the nature of mathematics. Then we shall shift to the applied point of view and spend the last three lectures looking at the mathematics of nature.

What is the nature of mathematics? Well, it consists of theorems and proofs. So we shall choose a few elegant little theorems and prove them. We will choose little theorems, because little ones are easy to state, and short to prove. But we will also select elegant ones, whose proofs are subtle and surprising, and were something of a triumph to the mathematicians who first discovered or invented them. We will choose theorems that capture the quintessence of a subject and epitomise its style, theorems in which a mathematician can see encapsulated in microcosm a great range of ideas and results; just as a literary critic might see the greatness of Shakespeare captured in a single sonnet, or an art historian might see the renaissance captured in a single painting by Massaccio. And to help our understanding we shall use many diagrams and pictures, because visual intuition is fundamental to our understanding of mathematics; it enables us to see things as a whole. Of course for proofs and calculations we sometimes have to use numbers and symbols and other mathematical tools, but often the technical details of these tools prevent us from seeing the wood for the trees, and it is then that our visual intuition comes to the rescue by helping us to perceive the overall pattern, even though the original problem may not have been anything to do with geometry. That is why the lectures are called mathematics into pictures, meaning pictures into understanding.

There have been three interesting features of the Royal Institution Christmas Lectures. The first is the number of people, in all walks of life, who say that their first realisation of the interest of science came from their attending, as children, a Christmas Lecture at the Institution. Almost invariably they say not that

'The lecturer told us so and so' but 'We were *shown* so and so.' It is the experiment that creates the vivid and lasting impression. In an interview given to the *Glasgow Herald*, March 1991, the Nobel prizewinner (and Honorary Member of the Royal Institution), Dorothy Hodgkin remarked that her attendance as a teenager at Sir William Bragg's Christmas Lectures in 1923 'Concerning the Nature of Things', convinced her that she should become a scientist.

The second feature about the Christmas Lectures is the number of popular scientific books which have been based on them. The lecturer is generally invited afterwards to put the Lectures into book form and many — some fifty in all — of the best popular science books published in the United Kingdom in the last 130 years have originated in this way. The third is the impact of television. Between Christmas and the New Year, when the Lectures are now broadcast, an aggregate viewing figure of some six to eight million is reached. Subsequent, repeat broadcasts throughout the year add a further four to five million. Moreover, when the abbreviated versions of the Christmas Lectures are shown in Japan, excerpts from them are televised from Tokyo, thereby swelling the number of children reached from the theatre of the Royal Institution. All this would have pleased Faraday.

It is interesting to contrast the number of children reached by Faraday during the nineteen series of Christmas Lectures that he presented with the number reached by one of the United Kingdom's most successful lecturers to schoolchildren in recent decades, Professor Charles Taylor,[3] who has given two enormously successful series of Christmas Lectures on exploring music. Faraday's aggregate (theatre) audience is estimated to have been 85,000. Taylor's theatre audience at the Royal Institution was close to 5,500; but his viewing audience well in excess of 20 million.

[Charles Taylor has probably presented more lecture-demonstrations to children (aged from seven to eighteen) than anyone else in Europe. Regularly at the Royal Institution for the past twenty years he has addressed approximately one thousand (of the thirty thousand) children that throng every year to its theatre to be inspired, educated and entertained in ways of which Faraday would have approved.]

Figure 63 Sir Lawrence Bragg giving the Christmas Lectures in 1961 on 'Electricity'. (From a painting by Cuneo, that hangs in the Royal Institution.)

Figure 64 A scene from the Christmas Lectures 1976-7 given by Sir George Porter with HRH The Duke of Kent sitting in the front row flanked (on his left) by HRH Prince Andrew and the Earl of St Andrews (right).

'They Have Their Exits and Their Entrances'

Lecturers who set out in the Christmas series to capture the hearts and reach the minds of young children tend, in the words of a master in the art, Sir Lawrence Bragg, to choose their experiments rather more for their intrinsic 'thrillingness' than for basic pedagogic content. Discourse speakers on the other hand, pay rather less attention to such considerations. The styles adopted by some two thousand speakers over a 165-year

203

period have varied enormously. Recognizing that a lecture is made or marred in the first ten minutes, it is interesting to recall how, over the years, various strategies have been employed to arrest the attention of Royal Institution audiences: how, too, speakers have drawn their subject to a close. The excerpts given below, taken, except where otherwise stated, from the *Proceedings of the Royal Institution*, have also been chosen for their intrinsic merits, as well as to illustrate the range of topics discussed and the state of knowledge pertaining to those topics at the time in question.

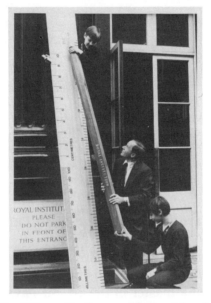

Figure 65 When Professor Philip Morrison of the Massachusetts Institute of Technology gave the Christmas Lectures on Gulliver's Laws the two sons of the then Director helped him handle a pencil and a foot ruler twelve times their normal size for use at the Lectures.

On **25 January 1878, The Right Honourable Thomas
H Huxley** (Fullerian Professor of Physiology at the
Royal Institution, and President of the Royal Society
1883-85) spoke on '**William Harvey**', the tercentenary
of whose birth was celebrated in 1878. T H Huxley is
remembered as 'Darwin's Bulldog', but he was a great
scientist in his own right. A self-made man who had
schooling only from the age of eight to ten, he taught
himself five foreign languages and later became an
assistant surgeon in the Royal Navy. A formidable
debator and brilliant stylist, he did much to dispel
religious and obscurantist opposition to the theory of
evolution. He began his Discourse on Harvey, the
tercentenary of whose birth was celebrated in 1878, as
follows:

> *Many opinions have been held respecting the exact
> nature and value of Harvey's contributions to the
> elucidation of the fundamental problem of the
> physiology of the higher animals; from those which
> deny him any merit at all — indeed, roundly charge him
> with the demerit of plagiarism — to those which
> enthrone him in a position of supreme honour among
> great discoverers of science. Nor has there been less
> controversy as to the method by which Harvey obtained
> the results which have made his name famous. I think
> it is desirable that no obscurity should hang around
> these questions; and I add my mite to the store of
> disquisitions on Harvey which this year is likely to
> bring forth, in the hope that it may help to throw light
> upon several points about which darkness has
> accumulated, partly by accident and partly by design.*

Huxley ended on a note which is highly relevant to the
debate that now rages between vivisectionists and their
opponents:

Figure 66 T H Huxley who was Fullerian Professor of Physiology and Comparative Anatomy at the Royal Institution and later President of the Royal Society.

The fact is that neither in this, nor in any, physiological problem can mere deductive reasoning from dead structure tell us what part that structure plays when it is a living component of a living body. Physiology attempts to discover the laws of vital activity, and these laws are obviously ascertainable only by observation and experiment upon living things.

In the case of the circulation of the blood, as in that of all other great physiological doctrines, take away the truths which have been learned by observation and experiment on living structures, and the whole fabric crumbles away. Galen, Columbus, Harvey, were all great vivisectors. And the final ocular demonstration of the circulation of the blood by Malpighi, seven years after Harvey's death — the keystone of the fabric he raised — involved an experiment on a living frog.

The novelist, journalist, scientific visionary and encyclopaedist **H G Wells** (1866-1946) when he presented his Discourse on **'The Discovery of the Future'** on 24 January 1902, was preoccupied with man's place in an inhospitable but ultimately conquerable universe, ideas which he was later to develop extensively in his novels and polemical writings. He began as follows:

It will lead into my subject most conveniently to contrast and separate two divergent types of mind, types which are to be distinguished chiefly by their attitude towards time, and more particularly by the relative importance they attach and the relative amount of thought they give to the future of things.

The first of these two types of mind — and it is, I think, the predominant type, the type of the majority of living people — is that which seems scarcely to think of the future at all, which regards it as a sort of black non-existence upon which the advancing present will presently write events. The second type, which is, I think, a more modern and much less abundant type of mind, thinks constantly, and by preference, of things to come, and of present things mainly in relation to the results that must arise from them. The former type of mind, when one gets it in its purity, is retrospective in habit, and it interprets the things of the present, and gives value to this and denies it to that, entirely with relation to the past. The latter type of mind is constructive in habit; it interprets the things of the present and gives value to this or that, entirely in relation to things designed or foreseen.

Later, in outlining his views on the scientific method he proclaimed:

But you will know that the essential thing in the scientific process is not the collection of facts, but the analysis of facts: facts are the raw material, and not the substance, of science; it is analysis that has given us all ordered knowledge; and you know that the aim, and the test, and the justification of the scientific process is not a marketable conjuring trick, but prophecy. Until a scientific theory yields confident forecasts, you know it is unsound and tentative; it is mere theorising, as evanescent as art talk, or the phantoms politicians talk about. The splendid body of gravitational astronomy, for example establishes itself upon the certain forecast of stellar movements, and you would absolutely refuse to believe its amazing assertions if it were not for these same unerring forecasts. The whole body of medical science aims, and claims the ability, to diagnose. Meteorology constantly and persistently aims at prophecy, and it will never stand in a place of honour until it can certainly foretell. The chemist forecasts elements before he meets them — it is very properly his boast; and the splendid manner in which the mind of Clerk Maxwell reached in front of all experiment, and foretold those things that Marconi has materialised, is familiar to us all. All applied mathematics resolves into computation to foretell things which otherwise can only be determined by trial. Even in so unscientific a science as political economy there have been forecasts.

'There may arise', said Wells, 'new animals to prey upon us by land and sea, and there may come some drug or wrecking madness into the minds of men.' He continued:

> *And finally, there is the reasonable certainty that this sun of ours must some day radiate itself towards extinction; that at least must happen; it will grow cooler and cooler, and its planets will rotate ever more sluggishly, until some day this earth of ours, tideless and slow moving, will be dead and frozen, and all that has lived upon it will be frozen out and done with. There surely man must end. That of all such nightmares is the most insistently convincing. And yet one doesn't believe it. At least I do not. And I do not believe in these things, because I have come to believe in certain other things, in the coherency and purpose of the world and in the greatness of human destiny. Worlds may freeze and suns may perish, but I believe that there stirs something within us now that can never die again.*

He culminated his Discourse in visionary mood:

> *We are in the beginning of the greatest change that humanity has ever undergone. There is no shock, no epoch-making incident; but then there is no shock at a cloudy daybreak. At no point can we say, 'Here it commences — and now, last minute was night and this is morning.' But insensibly we are in the day. If we care to look, we can foresee growing knowledge, growing order, and presently a deliberate improvement of the blood and character of the race. And what we can see and imagine gives us a measure and gives us faith for what surpasses the imagination.*
>
> *It is possible to believe that all the past is but the beginning of a beginning, and that all that is and has been is but the twilight of the dawn. It is possible to believe that all that the human mind has ever accomplished is but the dream before the awakening.*
>
> *We cannot see, there is no need for us to see, what this world will be like when the day has fully come. We are*

creatures of the twilight. But it is out of our race and lineage that minds will spring that will reach back to us in our littleness to know us better than we know ourselves, and that will reach forward fearlessly to comprehend this future that defeats our eyes. All this world is heavy with the promise of greater things, and day will come — one day in the unending succession of days — when beings, beings who are now latent in our thoughts and hidden in our loins, will stand upon this earth as one stands upon a footstool, and laugh, and reach out their hands amidst the stars.

Echoes and reflections of some of the thoughts and vistas conceived by Wells, along with totally new ones from realms unimaginable in his day, have been heard and seen at the theatre of the Royal Institution throughout the succeeding decades of this century.

Notes

1. By 'Films for the Humanities and Sciences', P.O. Box 2053, Princetown, NJ 08543-2053, USA.

2. 1848-49.

3. Charles A Taylor (1922-), formerly Professor of Physics, University College of South Wales and Monmouthshire, Cardiff (1965-83); Professor of Physics at the Royal Institution (1977-89); and first recipient (1986) of the Michael Faraday Medal awarded by the Royal Society for the popularization of science.

Epilogue

At the British Association meeting in 1866 a committee, which included John Tyndall and T H Huxley, was set up to consider the place of science teaching in schools and eventually through their representation the Government of the day was prevailed upon to set up a Royal Commission in 1870. Giving evidence to that Commission the great American scientist, first Director of The Smithsonian Institute and later President of the US National Academy, Joseph Henry, whose career paralleled that of Faraday in uncanny ways — including their emergence from humble origins and their interest in and discovery of electromagnetic induction — had this to say:

> *I have always looked upon the Royal Institution as a model establishment, doing honour to England, and producing an immense effect upon the world. More light has issued from that establishment in proportion to its means, than perhaps from any other on the face of the earth.*

The reader must judge to what extent this statement is still justified in 1991.

Appendix I

Faraday's Discourses & Scientific Publications, 1832-1834

Discourses:

1832

27 January.	Dr Johnson's Remarks on the Reproductive Power of Planariae.
17 February.	Recent experimental investigation of Volta-electric or Magnetic-electric induction.
2 March.	Magnetic-electric induction and the explanation it affords of Arago's phenomena of magnetism exhibited by moving metals.
30 March.	Evolution of Electricity naturally and artificially by the inductive action of the Earth's magnetism.
18 May.	The crispations of fluids lying on vibratory surfaces.
4 June.	Mordan's Machinery for Manufacturing Bramah's Locks.

1833

1 February.	The Identity of Electricity derived from different sources.
22 February.	The Practical Prevention of Dry Rot in Timber.
1 March.	An Investigation of the velocity and nature of the Electric Spark of Light.
29 March.	Mr Brunel's new mode of constructing arches.
3 May.	The mutual relations of lime, carbonic acid and water.

1834

24 January.	The Power of Platina and other solid substances to determine the combination of gaseous bodies.
14 February.	The Principle and Action of Ericson's Caloric Engine.
7 March.	Electro-chemical Decomposition.
11 April.	The Definite Action of Electricity.
24 May.	On a New Law of Electric Conduction.
13 June.	New applications of the products of Caoutchouc or Indian Rubber.

Publications

1832

1. Experimental researches in electricity. 2nd series. 5. Terrestrial magneto-electric induction. 6. Force and direction of magneto-electric induction generally. *Phil. Trans.* 1832: 163-94 (Bakerian lecture, read Jan. 12.)

2. On the Planariae. *Phil. Mag.* 1832, 11: 299.

3. On the first two parts of his recent researches in electricity: volta-electric induction, and magneto-electric induction. *Phil. Mag.* 1832, 11: 300-1.

4. On the explanation of Arago's phenomena of moving metals by magneto-electric induction. *Phil. Mag.* 1832, 11: 462.

5. Letter deposited at Royal Society on a theory of progressive magnetism. *Wireless World*, 1938, 42: 400-1.

6. Terrestrial magneto-electric induction, *Phil. Mag.* 1832, 11: 465-66.

7. Crispations of fluids, *Phil. Mag.* 1832, 1: 74.

8. On Mordan's apparatus for manufacturing Bramah's locks, *Phil. Mag.* 1832, 1: 75.

9. Syllabus of a course of five lectures upon some points of domestic chemical philosophy ... 2nd June-30th June, Report of last lecture in *Lit. Gaz.* 1832: 425.

10. On the new fowling piece of Wilkinson and Moser, *Lit. Gaz.* 832: 378.

11. On the electro-motive force of magnetism. By Signori and Antinori; from the *Antologia*, no. 131: with notes by Michael Faraday. *Phil. Mag.* 1832, 11: 402-13.

12. New experiments relative to the action of magnetism on electro-dynamic spirals, and a description of a new electro-motive battery. By Signor Salvatore dal Negro; with notes by Michael Faraday. *Phil. Mag.* 1832, 1: 45-49.

13. Account of an experiment in which chemical decomposition has been effected by the induced magneto-electric current. By P.M.; preceded by a letter from Michael Faraday. *Phil. Mag.* 1832, 1: 161-62.

14. Lettre à M. Gay-Lussac sur l'électro-magnétisme, *Ann. Chim.* 1832, 51: 404-34.

1833

15. Six lectures on chemistry, *Lit. Gaz.* 1833: 11.

16. Experimental researches in chemistry. 3rd series. 7. Identity of electricities derived from different sources. 8. Relation by measure of common and voltaic electricity, *Phil. Trans.* 1833: 23-54 (read Jan. 10 and 17).

17. On the identity of electricity derived from different sources [title from ms. notes], *Phil. Mag.* 1833, 2: 312.

18. Report of Committee appointed by the Royal Society to examine the Proof standard for alcohol, F. a member of this Committee and presented its report on Feb. 12, 1833.

19. On the practical prevention of dry rot in timber; being the substance of a lecture delivered by Professor Faraday at the Royal Institution, February 22, 1833. With observations, etc. London, J and C Adlard,

       printers, Bartholomew Close, 1833. Reports appeared in *Phil. Mag.* 1833, **2**: 313-14; *Lit. Gaz.* 1833: 136, and *Athenanaeum* 1833: 139.

20. Investigation of the velocity and other properties of electric discharges [title from ms. notes], *Lit. Gaz.* 1833: 152.

21. Address delivered at the commemoration of the centenary of the birth of Dr Priestley, *Phil. Mag.* 1833, **2**: 390-91.

22. On Mr Brunel's new mode of building arches [title from ms. notes], *Lit. Gaz.* 1833: 217.

23. On the mutual relations of lime, carbonic acid and water, *Lond. Med. Gaz.* 1832-33, **12**: 191-92.

24. Experimental researches in electricity. 4th series. 9. On a new law of electric conduction. 10. On conducting power generally. *Phil. Trans.* 1833: 507-22 (read May 23).

25. On a new law of electric conduction [title from ms. notes], *Lit. Gaz.* 1833: 345.

26. Experimental researches in electricity. 5th series. 11. On electro-chemical decomposition. *Phil. Trans.* 1833: 675-710 (read June 20).

27. Notice of a means of preparing the organs of respiration, so as considerably to extend the time of holding the breath; with remarks on its application, in cases in which it is required to enter an irrespirable atmosphere, and on the precautions necessary to be observed in such cases, *Phil. Mag.* 1833, **3**: 241-44.

1834

28. Experimental researches in electricity. 6th series. 12. On the power of metals and other solids to induce the combination of gaseous bodies. *Phil. Trans.* 1834: 55-76 (read Jan. 11).

29. Experimental researches in electricity. 7th series. 11. On electro-chemical decomposition (continued). 13. On the absolute quantity of electricity associated with the particles or atoms of matter. *Phil. Trans.* 1834: 77-122 (read Jan. 23, Feb. 6 and 13).

30. On the power of the platina and other solid substances to determine the combination of gaseous bodies, *Athenaeum* 1834: 90-91.

31. On the principle and action of Ericsson's caloric engine, *Phil. Mag.* 1834, **4**: 296.

32. On electro-chemical decomposition, *Athenaeum* 1834: 209.

33. On the definite action of electricity, *Athenaeum* 1834: 296.

34. Experimental researches in electricity. 8th series. 14. On the electricity of the voltaic pile; its source, quantity, intensity, and general characters. *Phil. Mag.* 1834: 425-70 (read June 5).

35. On new applications of the products of distilled caoutchouc [title from ms. notes], *Lit. Gaz.* 1834: 435.

36. On the magneto-electric spark and shock, and on a peculiar condition of electric and magneto-electric induction, *Phil. Mag.* 1834, **5**: 349-54.

37. Additional observations respecting the magneto-electric spark and shock, *Phil. Mag.* 1834, **5**: 444-45.

38. Report from Select Committee on Metropolis Sewers, *Parl. Pap.* 1834 (584) xv.

Appendix II

Learned Societies to which Faraday was Elected

1823 Corresponding member of the Academy of Sciences, Paris.
Corresponding member of the Accademia dei Georgofili, Florence.
Honorary member of the Cambridge Philosophical Society.
Honorary member of the British Institution.
1824 Fellow of the Royal Society.
Honorary member of the Cambrian Society, Swansea.
Fellow of the Geological Society.
1825 Member of the Royal Institution.
Corresponding member of the Society of Medical Chemists, Paris.
1826 Honorary member of the Westminster Medical Society.
1827 Correspondent of the Société Philomathique, Paris.
1828 Fellow of the Natural Society of Science, Heidelberg.
1829 Honorary member of the Society of Arts, Scotland.
1831 Honorary member of the Imperial Academy of Sciences, St Petersburg.
1832 Honorary member of the College of Pharmacy, Philadelphia.
Honorary member of the Chemical and Physical Society, Paris.
Fellow of the American Academy of Arts and Sciences, Boston.
Member of the Royal Society of Science, Copenhagen.
1833 Corresponding member of the Royal Academy of Sciences, Berlin.
Honorary member of the Hull Philosophical Society.
1834 Foreign corresponding member of the Academy of Sciences and Literature, Palermo.
1835 Corresponding member of the Royal Academy of Medicine, Paris.
Honorary member of the Royal Society, Edinburgh.
Honorary member of the Institution of British Architects.
Honorary member of the Physical Society, Frankfurt.
Honorary Fellow of the Medico-Chirurgical Society, London.
1836 Senator of the University of London.
Honorary member of the Society of Pharmacy, Lisbon.
Honorary member of the Sussex Royal Institution.

Foreign member of the Society of Sciences, Modena.
Foreign member of the Natural History Society, Basle.
1837 Honorary member of the Literary and Scientific Institution, Liverpool.
1838 Honorary member of the Institution of Civil Engineers.
Foreign member of the Royal Academy of Sciences, Stockholm.
1840 Member of the American Philosophical Society, Philadelphia.
Honorary member of the Hunterian Medical Society, Edinburgh.
1842 Foreign Associate of the Royal Academy of Sciences, Berlin.
1843 Honorary member of the Literary and Philosophical Society, Manchester.
Honorary member of the Useful Knowledge Society, Aix-la-Chapelle.
1844 Foreign Associate of the Academy of Sciences, Paris.
Honorary member of the Sheffield Scientific Society.
1845 Corresponding member of the National Institute, Washington.
Corresponding member of the Société d'Encouragement, Paris.
1846 Honorary member of the Society of Sciences, Vard.
1847 Member of the Academy of Sciences, Bologna.
Foreign Associate of the Royal Academy of Sciences of Belgium.
Fellow of the Royal Bavarian Academy of Sciences, Munich.
Correspondent of the Academy of Natural Sciences, Philadelphia.
1848 Foreign honorary member of the Imperial Academy of Sciences, Vienna.
1849 Honorary member, first class, of the Institut Royal des Pays Bas.
Foreign correspondent of the Institute, Madrid.
1850 Corresponding Associate of the Accademia Pontificia, Rome.
Foreign Associate of the Academy of Sciences, Haarlem.
1851 Member of the Royal Academy of Sciences, The Hague.
Corresponding member of the Batavian Society of Experimental Philosophy, Rotterdam.
Fellow of the Royal Society of Sciences, Upsala.
1853 Foreign Associate of the Royal Academy of Sciences, Turin.
Honorary member of the Royal Society of Arts and Sciences, Mauritius.
1854 Corresponding Associate of the Royal Academy of Sciences, Naples.
1855 Honorary member of the Imperial Society of Naturalists, Moscow.
Corresponding Associate of the Imperial Institute of Sciences of Lombardy.
1856 Corresponding member of the Netherlands' Society of Sciences, Batavia.
Member of the Imperial Royal Institute, Padua.
1857 Member of the Institute of Breslau.
Corresponding Associate of the Institute of Sciences, Venice.
Member of the Imperial Academy, Breslau.
1858 Corresponding member of the Hungarian Academy of Sciences, Pesth.
1860 Foreign Associate of the Academy of Sciences, Pesth.
Honorary member of the Philosophical Society, Glasgow.
1861 Honorary member of the Medical Society, Edinburgh.
1863 Foreign Associate of the Imperial Academy of Medicine, Paris.
1864 Foreign Associate of the Royal Academy of Sciences, Naples.

Appendix III

Faraday's Friday Evening Discourses, 1835 onwards

DATE	TOPIC	ATTENDANCE
23-1-1835	Melloni's Recent Discoveries in Radiant Heat.	400
6-2-1835	The Induction of Electric Currents.	460
27-3-1835	The Manufacture of Pens from Quills and Steel Illustrated by Modern Machinery.	528
15-5-1835	The Condition and Use of the Tympanum in the Ear.	405
22-1-1836	Silicified Plants and Fossils.	416
19-2-1836	The Magnetism of Metals as a General Character.	674
29-4-1836	Plumbago and the Manufacture of Pencils from it with Modern Machinery.	662
10-6-1836	Considerations Respecting the Nature of Chemical Elements.	646
20-1-1837	Mosoth's Reference of Electrical Attraction, the Attraction of Aggregation and the Attraction of Gravity to One Cause.	467
17-2-1837	Dr Marshall Hall's Reflex Function of the Spinal Marrow. Mr Cowper's Printing Press.	480
17-3-1837	Mr de la Rue's Mode of Applying Sulphate of Copper to the Exaltation of the Powers of a Common Voltaic Battery.	675
28-4-1837	A Peculiar Condition of Iron in Relation to its Chemical Affinity in its Electro-Motive Force.	582
9-6-1837	Early Arts: The Bow and Arrow.	583
19-1-1838	Electrical Induction.	382
28-2-1838	The Atmosphere of This and of Other Planets.	435
6-4-1838	Mr Ward's Mode of Growing and Preserving Plants in Limited Atmospheres.	625
18-5-1838	The Solid, Liquid and Gaseous State of Carbonic Acid Illustrated by Philosier's Apparatus From Professor Graham.	602
8-6-1838	The Relation of Electric Induction and Insulation.	714
18-1-1839	On the Gymnotics of the Torpedo.	554

15-2-1839 On Gunney's Oxy-Oil Lamp. 705
22-3-1839 On Airy's Correction of the Ship's Compass in an Iron Vessel. 656
10-5-1839 Some General Remarks on Flame. 540
7-6-1839 Hulmandell's Mode of Producing Designs and Patterns on 552
    Metallic Surfaces.
24-1-1840 On Voltaic Precipitations. 472
7-2-1840 On a Particular Relation to the Condensible Gases to Steam. 204
8-5-1840 On the Origin of Electricity in the Voltaic Pile. 675
15-4-1842 Conduction of Electricity in Lightning Rods. 773
10-6-1842 The Principles and Practice of Hulmandell's Lithotint. 798
20-1-1843 Some Phenomena of Electric Induction. 555
7-4-1843 The Ventilation of Lamp Burners. 998
9-6-1843 The Electricity of Steam. 836
19-1-1844 Speculations Touching Electric Conduction and the Nature of 732
    Matter.
7-6-1844 Recent Improvements in the Manufacture and Silvering of 866
    Mirrors.
17-1-1845 The Condition and Ventilation of the Coal-Mine Goat. 643
31-1-1845 The Liquefaction and Solidification of Bodies Usually Gases. 861
25-4-1845 Anastatic Printing. 873
30-5-1845 The Artesian Well and Water. 734
23-1-1846 Magnetism and Light. 1003
6-3-1846 The Magnetic Conditions of Matter. 1000
3-4-1846 Mr Wheatstone's Electro-Magnetic Chronoscope. 706
12-6-1846 The Cohesive Force of Water. 546
22-1-1847 Gun-Powder. 658
26-3-1847 Mr Barry's Mode of Ventilating the New House of Lords. 695
11-6-1847 The Steam-Jet. 538
14-4-1848 The Diamagnetic Condition of Flame and Gases. 909
26-5-1848 On Two Recent Inventions of Artificial Stone; One Entirely 367
    Siliceous and Available for Architectural Decorations. The
    Other a Breccia Formed in Moulds From Fragments of Any
    Kind of Stone and Applicable to All Kinds of Building and
    the Pipes, Sewers and Heavy Underwater Works.
16-6-1848 The Conversion of Diamond into Coke. 643
26-1-1849 The Crystalline Polarity of Bismuth and Other Bodies and 677
    its Relation to the Magnetic Force.
26-2-1849 The Diamagnetic and Magne-Crystallic Condition of Bodies. 331
    HRH Prince Albert present.
30-3-1849 Plucker's Repulsion of the Optic Axes of Crystals by the 692
    Magnetic Poles.
1-6-1849 Envelope Machinery. 619
1-2-1850 The Electricity of the Air. 806
8-6-1850 Certain Conditions of Freezing Water. 888
24-1-1851 On the Magnetic Character and Relation of Oxygen and 813
    Nitrogen.
11-4-1851 On Atmospheric Magnetism. 1028
13-6-1851 On Schönbein's Ozone. 692
11-6-1852 On the Physical Lines of Magnetic Force. 895
23-6-1852 On the Lines of Magnetic Force. 670
21-1-1853 Observations on the Magnetic Force. 830

10-6-1853	MM Boussingault, Frémy, Bequerel etc. on Oxygen.	836
20-1-1854	On Electric Induction — Associated Causes of Current and Static Effects.	762
9-6-1854	On Magnetic Hypotheses.	806
19-1-1855	On Some Points of Magnetic Philosophy.	576
25-5-1855	On Electric Conduction.	562
8-6-1855	On Ruhmkorff's Induction Apparatus.	663
22-2-1856	On Certain Magnetic Actions and Affections.	903
13-6-1856	On M. Petitjean's Process for Silvering Glass: Some Observations on Divided Gold.	680
27-2-1857	On the Conservation of Force.	871
12-6-1857	On the Relations of Gold to Light.	735
12-2-1858	Remarks on Static Induction.	796
11-6-1858	On Wheatstone's Electric Telegraph in Relation to Science (Being an Argument in Favour of the Full Recognition of Science as a Branch of Education).	753
25-2-1859	On Schönbein's Ozone and Antozone.	876
9-3-1860	On Lighthouse Illumination — the Electric Light.	797
8-6-1860	On the Electric Silk-Loom.	698
22-2-1861	On Platinum.	883
	(Printed and bound with *Chemical History of a Candle*.)	
3-5-1861	On Mr Warren de la Rue's Photographic Eclipse Results.	783
20-6-1862	On Gas Furnaces.	812

Appendix IV

Discourses Arranged by Faraday (before 1862)

DATE	LECTURER	TOPIC	ATTENDANCE
24-2-1832	Marquis Moscati	The genius of the Extemporaneous Poets and on the Art of Improvisation by an Italian Improvisator.	418
9-3-1832	George Foggo	The Causes of the Excellence of Grecian Art.	355
13-4-1832	Marshall Hall	The Laws which Govern the Mutual Relation of Respiration and Irritability.	320
21-3-1834	William Varlo Hellyer	A Day at Pompeii.	529
25-4-1834	John Davidson	The Pyramids of Egypt.	720
2-5-1834	Dionysius Lardner	Babbage's Calculating Machinery.	708
9-5-1834	John Dalton	On the Atomic Theory of Vapours.	594
13-2-1835	John Landseer	A Sculptured, Historical Monument Lately Brought From Phoenecia by Mr Joseph Bonami and Now in the Possession of Lord Prudhoe.	360
10-4-1835	Dionysius Lardner	Notice of Halley's Comet.	820
6-5-1836	John Frederick Daniell	A New and Constant Voltaic Battery.	503
27-5-1836	Thomas Joseph Pettigrew	The Opening of an Egyptian Mummy.	810
3-6-1836	Richard Beamish	The Present State and Prospects of the Thames Tunnel.	426
12-5-1837	Gideon Algernon Mantell	The Iguanadon and Other Fossil Remains Discovered in the Strata of Filgate Forest.	471

19-5-1837	Monsieur De La Rue	The History and Manufacture of Playing Cards.	328
25-5-1838	John Landseer	The Astronomy of the Book of Job.	323
5-2-1841	Samuel Birch	Hieroglyphics of the Egyptians.	348
19-2-1841	James Tennant	Ornamental Stones used in Jewellery.	354
19-3-1841	John Joseph Cooper	Elkingtons' New Mode of Plating and Gilding.	455
26-3-1841	J F Goddard	The Application of the Daguerrotype to the Taking of Likenesses from Life.	621
7-5-1841	Thomas G Griffiths	The Manufacture of Soda Water.	188
28-5-1841	Sir Richard Owen	The Method of Investigating Fossil Remains.	304
10-2-1843	Sir William Robert Grove	The Gaseous Voltaic Pile.	458
17-3-1843	Owen Jones	Moorish Architecture as Illustrated by the Alhambra.	583
6-6-1845	Sir Roderick Impey Murchison	Russia and the Ural Mountains.	381
8-5-1846	John Scott Russell	The Application of Certain Laws of Sound to the Construction of Buildings.	231
30-4-1847	Sir Charles Lyell	The Age of the Volcanoes of Auvergne as Determined by the Remains of Successive Groups of Land-Quadrupeds.	383
7-5-1847	Tom Taylor (Editor of *Punch*)	The Saxon Epic — Beowulf.	222
4-2-1848	Sir Charles Lyell	The Fossil Footmarks of a Reptile in the Coal Formations of the Alleghany Mountains.	389
7-4-1848	Rev Baden Powell	Shooting Stars and Their Connection with the Solar System.	410
2-6-1848	John Scott Russell	On the Tidewave Principle Applied to the Construction of Ships.	235
19-1-1849	Rev William Whewell	The Idea of Polarity.	404
8-2-1850	Edward Cowper	The Conway and Menai Tubular Bridges.	605
2-5-1851	Sir George Biddell Airey, PRS	On the Total Solar Eclipse of July 28, 1851.	610
30-5-1851	Sir Henry Creswicke Rawlinson	A Few Words on Babylon and Nineveh.	647
6-6-1851	Alexander William Williamson	Suggestions for the Dynamics of Chemistry Derived From the Theory of Etherification.	296
8-3-1852	Rev William Taylor	Observations on Different Modes of Educating the Blind.	310
30-4-1852	Rt Hon Thomas Henry Huxley	On Animal Individuality.	323
18-2-1853	George Gabriel Stokes	On the Change of Refrangibility of Light and the Exhibition Thereby of the Chemical Rays.	450

6-5-1853	Lyon Playfair (1st Baron Playfair)	On the Food of Man Under Different Conditions of Age and Employment.	498
3-6-1853	John Tyndall	On Some of the Eruptive Phenomena of Iceland.	510
24-2-1854	Henry Bence Jones	On the Acidity, Sweetness and Strength of Different Wines.	460
18-5-1855	Sir James Philip Lacaita	On Dante and the 'Divina Commedia'.	410
15-6-1855	Sir Henry Creswicke Rawlinson	On the Results of the Excavations in Assyria and Babylonia.	906
15-2-1856	Rt Hon Thomas Henry Huxley	On Natural History, as Knowledge, Discipline and Power.	447
29-2-1856	William Thomson (Lord Kelvin)	On the Origin and Transportations of Motive Power.	449
11-4-1856	Sir Charles William Siemens	On a Regenerative Steam-Engine.	310
30-1-1857	Rev Frederick Denison Maurice	Milton Considered as a Schoolmaster.	422
5-3-1858	Charles Piazzi Smyth	Account of the Astronomical Experiment of 1856 on the Peak of Tenerife.	521
28-5-1858	Sir Edward Frankland	On the Production of Organic Bodies Without the Agency of Vitality.	328
4-2-1859	Sir Richard Owen	On the Gorilla.	553
10-2-1860	Rt Hon Thomas Henry Huxley	On Species and Races and Their Origin.	516
18-5-1860	William Thomson (Lord Kelvin)	On Atmospheric Electricity.	322
8-2-1861	Rt Hon Thomas Henry Huxley	On the Nature of the Earliest Stages of the Development of Animals.	461
12-4-1861	Hermann Ludwig Ferdinand von Helmholtz	On the Application of the Law of the Conservation of Force to Organic Matter.	332
19-4-1861	John Ruskin	On Tree Twigs.	805
17-5-1861	James Clerk Maxwell	On the Theory of Three Primary Colours.	408
17-1-1862	John Tyndall	On the Absorption and Radiation of Heat by Gaseous Matter.	482
21-3-1862	Sir Frederick Augustas Abel	On Some of the Causes, Effects and Military Applications of Explosives.	364

Appendix V

Part of The Royal Institution Lecture Calendar, 1933

FEBRUARY			LECTURE HOUR
2	Thurs	*J B S Haldane* — Recent Advances in Genetics	5.15
3	Fri	*Cyril Norwood* — Use of the English Language	9
4	Sat	*L Binyon* — Oriental Painting	3
6	Mon	General Meeting	5
7	Tues	*J C M'Lennan* — Low Temperatures	5.15
9	Thurs	*J B S Haldane* — Recent Advances in Genetics	5.15
10	Fri	*A V Hill* — Physical Nature of the Nerve Impulse	9
11	Sat	*L Binyon* — Oriental Painting	3
14	Tues	*Sir William Bragg* — Analysis of Crystal Structure by X-Rays	5.15
16	Thurs	*A R Hinks* — Geography in the Public Service	5.15
17	Fri	*J Dover Wilson* — Plot of Hamlet	9
18	Sat	*Lord Rutherford* — Detection and Production of Swift Particles	3
21	Tues	*Sir William Bragg* — Analysis of Crystal Structure by X-Rays	5.15
23	Thurs	*A R Hinks* — Geography in the Public Service	5.15
24	Fri	*W A Bone* — Photographic Analysis of Explosion Flames	9
25	Sat	*Lord Rutherford* — Detection and Production of Swift Particles	3
28	Tues	*Sir William Bragg* — Analysis of Crystal Structure by X-Rays	5.15

Index

Académie des Sciences, 50, 62

Accademia in Florence, 19, 20

Adsorption, selective, 87, 89

Airy, Sir George (Astronomer Royal), 136

Albert, Edward, HRH Prince, 120

Albert, HRH Prince, 3, 120, 127, 160

Aluminium, discovery of, 31

Alloy steels, 23, 25

Ampère, André Marie, 46, 68

Analytical chemistry, 23

Arabian Nights, 118

Arago, Dominque, 19, 105

Arnold, Matthew, 136, 188

Arrhenius, Svante, 136, 188

Astbury, W T, a founder of molecular biology, 183, 190

Astronomy, 37, 176

Athenaeum, The, 3, 12, 33, 37

Attenborough, Sir David, 199

Austin, Alfred (Poet Laureate), 136

Australia and New Zealand Association for Advancement of Science, 140

Babbage, Charles, 46

Baird, Logie, 183

Baker, Samuel White (explorer), 137

Bakerian Lecture, 3, 6, 12, 78, 82, 122, 147

Balfour, A J (Prime Minister), 180, 190

Banks, Sir Joseph, 4, 7

Barkla, Charles, 137

Barnard, Sarah (wife of Faraday), 27, 118

Bateson, William, 137

Bavaria, Elector of, 4

Becquerel, Henri, 107, 137

Bedford, Duke of, 13

Bell, Alexander Graham, 137, 166, 167

Bence Jones, H, 74, 153

Berzelius, Jöns Jacob, 27, 34, 46, 92

Blunt, Anthony, 188

Bose, Jagadis Chunder, 137

Botanical Gardens, Kew, 154
Bragg, Sir Lawrence (W L), 115, 137, 183, 185, 192, 202, 203
Bragg, Sir William (W H), 115, 137, 183, 201
Brande, W T, 25, 34, 38
British Association, 123, 150, 153
British Museum, 3, 83
Buffon, Comte de, 128
Bunsen, R W, 92, 151, 191
Butler, Samuel, 128
Byron, Lord, 13

Cadogan, Sir John, 140
Cambridge, Philosophical Society, 130
Cambridge, Trinity College, 3, 46, 130
Candle, The Chemical History of a (M Faraday), 2, 95, 193-195
Cannizzaro, S, 186-187, 191
Cantor, G A, 131
Caroe, Gwendy, 93, 190
Catalysis, 23, 87
Cavendish Laboratory (Cambridge), 8, 49
Cavendish, Henry, 29, 43, 58, 61, 92, 93
Chemical Manipulations (M Faraday), 38, 95
Chemical Society, 150
Chladni plate, 140, 141, 156
Chlorine, 33

Christmas Lectures, 2, 37, 39, 60, 65, 174, 192-193, 200
City Philosophical Society, 24, 28
Clarendon Laboratory (Oxford), 8
Clathrates, 31
Coleridge, S T, 7, 12
Colloidal metals, 23, 78-81, 93
Commonplace book (of Faraday's), 15
Concept of field, 43
Concerning the Nature of Things (W H Bragg), 201
Constable, John, 33, 134, 135
Cooke, W F, 39, 124
Correlation of Physical Forces (W R Grove), 122, 147
Coulometry, 52
Coutts, Baroness Burdett, 116, 130
Crimean War, 131
Critical temperature, 23, 33
Crookes, Sir William, 190, 195, 196
Curie, Madame (Marie), 176, 190
Curie, Pierre, 176
Cuvier, George, 19

Dalton, John, 61, 62, 151
Dana, J D, 191
Dance, Mr, 16
Daniell, John Frederick, 66

Danish Society for the Promotion of Scientific Knowledge, 31
Darwin's bulldog, 205
Darwin, Charles, 3
Davies, Paul, 77, 78
Davy, Sir Humphry, 2, 4, 5, 7, 9-13, 15-20, 30, 31, 33, 37, 40, 62, 90, 137
    agricultural chemistry, 12
    and the Greeks, 12
    bleaching cloth, 11
    cathodic protection, 11
    dealings with Faraday, 37
    discovery of barium, boron, calcium, magnesium, potassium, sodium, strontium, 8
    electric arc, 11
    London Zoo, 12
    miners' safety lamp, 11, 23
    paintings on ceramics, 11
    poetic flair, 12
    tanning of leather, 11
Davy, John, 10
De La Rue, Warren, 139
de la Rive, Auguste, 20, 105, 123, 156
de la Rive, Gustave, 20
Debye, Peter, 64
Dewar, Sir James, 19, 174-177
Diamagnetism, 23-24, 65, 70
Dickens, Charles, 3
Dielectric constant, 23, 61, 64
Discovery of the Future (H G Wells), 207-210
Dumas, J B, 186
Dye industry, 35

Dynamo, 40

Eddington, Sir Arthur, 76
Eddy currents, 71
Edison, T Alvar, 165
Einstein, Albert, 1
Electric motor, 29
Electrical discharges in gases, 56
Electrical industry, birth of, 40
Electricity, identity of different kinds of, 24, 52, 53
Electricity and gravity, 24, 43, 74-78
Electroanalysis, 52
Electrochemistry, language of, 50
Electrolysis, laws of, 23, 45
Electromagnetic induction, 24, 40-41, 43
Electromagnetic rotations, 24, 29, 32, 40
Electroplating, electrogilding and electroforming, 50-52
Electrostatics, 24, 58
Elgin Marbles, 3
Eliot, George (*Middlemarch*), 12
Emerson, R Waldo, 136, 188
Encyclopaedia Britannica, 15, 118
Equivalence of voltaic, static, thermal and animal electricity, 23

Experimental Researches in Chemistry and Physics (Faraday's compilation of own papers), 84, 95
Experimental Researches in Electricity and Magnetism (M Faraday), 71, 95

Faraday as Discoverer (J Tyndall), 129, 157
Faraday, Michael, 15, 36, 197
    analysis of genius, 128
    analytical chemist, 23, 25
    and mathematics, 1-2
    Arabian Nights, 118
    as errand boy and bookbinder, 15
    benzene, 23, 34-35
    Bibles, 117, 121, 131
    bicarburet of hydrogen (benzene), 23, 34-35
    brass nameplates, 119
    brother, Robert, 34
    cage, 24, 58, 59, 74
    clathrates, 31
    colloidal metals, 23, 78-81, 93
    condensers, 60, 62
    concept of field, 43
    continuity of state, 23
    conservation of forces, 101
    correspondence with W R Grove, 147-148

critical temperature, 23, 33
dark space, 56
Dean of St Paul's, 83, 117
diamagnetism, 23-24, 65, 70
Diary, excerpts from, 41, 44, 59, 88, 89, 96
dielectric work, 23, 58, 61, 64
discharge between conductors, 56
Discourses arranged by, 42, 138, 212, 217-22
discovery of new compounds, 26
divided metals, 78, 80, 81, 93
dynamo, 40
effect, 65, 69
electromagnetic rotation, 24, 29, 32, 40
electromagnetism, 24, 40
electrostatic work, 24, 58
exhaustion, 64
existence of atoms, 61, 62
first electric motor, 29
first paper, 24
founder of electrical industry, 40, 43
gas hydrates, 31
glass of optical quality, 23, 36, 82
Hofmann, views on, 191
Holding the interest of an audience, 99
induction of electric current, 100

last experimental work, 83

laws of electrolysis, 23, 45, 48

laws of electromagnetic induction, 43

learned societies to which he was elected, 215-6

letters to *The Times*, 127, 132

lighthouses, 60, 131

lightning, views on, 111-114

lines of force, 44, 47, 67-68, 130

liquefied chlorine and other gases, 23, 33

magnetochemistry, 23, 72

magneto-optics, 23, 65

naphthalene, 35

optimism, 102

organic compounds discovered by, 35

paramagnetism, 23-24

permittivity, 23, 61

polarization of molecules, 60, 63

principal contributions to chemical science, 23

principal contributions to physical science, 24

rotation of plane of polarization of light, 65-68

Royal Society, election to, 33

refusal to accept Presidency of, 124

superionic conductors, 23, 87

Scientific Advisor to Trinity House, 58

table-turning and the Spiritualists, 125, 127

thermistor effect, discovery of, 23, 88

thin films, 79

transformer, 40

vulcanization of rubber, 23, 38, 91

wife, Sarah Barnard, 27, 118

Faraday in Wales (Dafydd Tomos), 119

Fleming, Sir Alexander, 190

Forbes, J D, 46

Foucault, Léon, 71

Fox Talbot, W H, 3, 139, 152, 153

Frankland, Edward, 49, 92

Franklin, Benjamin, 52, 128

French Academy of Sciences, 50, 62

Fuel cell, 147-149

Fuller, John, 54, 57

Fullerian Professorship of Chemistry, 54, 175

Fused salt electrolytes, 23, 87

Gay-Lussac, Joseph, 19, 46

GEC, 43
Gell-Mann, Murray, 188
General Electric,
    Schenectady, 43
Genius of Faraday, an
    analysis, 128
Geological Society, 12
George III, King, 4
Gibbs, J Willard, 191
Gladstone, W E, 127, 168
Glasgow Herald, 201
Glass, production of optical
    grade, 36
Gold, colloidal, 78, 79, 81, 93
Gombrich, Sir Ernst, 134, 136
Gowing, Margaret, xii
Gravity and electricity, link
    between them, 24, 43,
    74-78
Greenhouse effect, 153
Grove, William Robert (later
    Sir and Judge), 110, 122,
    142-150, 153
Guest, John, 25
Gulliver's Laws, 199

Hale, George (astronomer),
    86, 176
Harvey, William, 205
Healey, Edna (*Lady
    Unknown*), 131
Helmholtz, Hermann von, 49,
    191
Henry, Joseph, 211
Herschel, J W H (Sir John),
    36, 46
Herschel, Sir William, 36
Hertz, Heinrich, 68

Hill, A V, 223
Hodgkin, Dorothy, 201
Hoffmann, Roald, 188
Hofmann, von A W, 92, 159-
    162, 191
Humboldt, Baron von, 19
Huxley, T H, 3, 168, 205-207,
    211

Indian Association for the
    Cultivation of Science,
    82
Induction of electric currents,
    100
Inverse square law, 43
Irving, Sir Henry, 175-6

Japan, Christmas Lectures in,
    193, 201
Jay, Sir Antony, 189
Jeans, Sir James, 197
John Innes (Horticultural
    Institute), 137
Jonson, Ben (on Shakespeare),
    3

Kelvin, Lord (William
    Thomson), 3, 68, 75
Kendrew, Sir John, 185
Kent, HRH Duke of, 203
Kerker, Milton, 93, 190
Keynes, J Maynard, 183
Kirchhoff, G R, 151
Krishnan, K S, 72

Lady Unknown (Edna
    Healey), 131

Last Words of a Sensitive Second-rate Poet (Owen Meredith), 128
Lavoisier, Anne, 5
Lavoisier, Antoine, 5, 11
Liebig, Justus von, 160
Life and Letters of Faraday, (Bence Jones), 74, 153
Liquefaction of gases, 23, 33
Lithography, 38
Lodge, Sir Oliver, 85, 86, 190
Lonsdale, Dame Kathleen, 72
Lorentz, Hendrik, 85, 188
Lorenz, Konrad, 188
Lowell, Percival (astronomer), 176
Lysozyme, structure solved in Royal Institution, 186

Magnetic anisotropy, 72
Magnetic character of oxygen, 107
Magnetochemistry, 23, 72
Magneto-optics, 23, 65
Marcet, Mrs (*Conversations on Chemistry*), 15
Marconi, Gugliemo, 163
Mathematics into Pictures, 199
Mauve and magenta dye-stuffs, 160, 162
Maxwell, J Clerk, 1, 22, 45, 65, 68-69, 73-75, 83, 85, 130, 180
McCabe, I M, 115, 189
Mead, Margaret, 188
Melbourne, Lord, 55-56
Mendeleeff, D I, 186-187, 191

Meredith, Owen, 128
Metallography, 23
Microcosm of London (Rowlandson), 14
Miners' Safety Lamp, 11, 23
Mond, Ludwig, 182
Morrison, Philip, 204
Murchison, Sir Rhoderick, 107, 111
Muybridge, Eadweard, 163, 167-173

Napoleon, prize, 12
his regard for science, 19
National Gallery, 3, 83
Nature of Mathematics, 199
NEC, 43
Newton, Isaac, 1, 29, 130

Ørsted, Hans Christian, 29-31, 40, 90, 122
Organic chemistry, 23, 26
Ostwald, Wilhelm, 62, 188

Paramagnetism, 23-24
Pasteur, Louis, 154, 191
Pauling, Linus, 188
Peel, Sir Robert, 55, 127
Perkin, William Henry, 160
Permittivity, 23, 61
Perrin, J B, 79, 94, 188
Perry, Geoffrey, 93
Perutz, Max, 185
Philips Electrical Industries, 43
Phillips, David C, 185
Photochemical preparations, 23

Piazzi Smyth, Charles
(Astronomer Royal for
Scotland), 142-144, 151,
189
Pippard, Sir Brian, xii
Pixii, Hippolyte, 50
Plasmas, chemistry and
physics of, 56
Pluto, prediction of its
existence, 176
Polarization of molecules, 60,
63
Popularization of science, 192
Porter, Sir George (later
Lord), 115, 140, 192, 203
Preece, Sir William Henry,
162-167
Prince of Wales, 168
Punch (cartoon), 125, 126

Quakers, 28
Quarterly Journal of Science,
25

Radium, Discourse on, 176
Raman effect, 81
Raman, Sir C V, 82
Ramsay, William, 180
Rayleigh, Lord (John
William Strutt), 174,
180-183, 190
Riebau, George, 15
Robinson, Sir Robert, 38
Roget, Mark (of *Thesaurus*
fame), 3, 57
Rosetta Stone, 6
Royal Military Academy,
Woolwich, 38, 46, 91,
150

Royal Society, 4, 12, 30, 33,
36, 67, 69, 82, 86, 122,
124, 136, 149, 150, 155,
190, 195
Rumford, Count (Benjamin
Thompson), 4, 6, 19, 155
Ruskin, John, 188
Rustless steels, 23, 25
Rutherford, Lord, 18, 129, 183,
188

Salam, Abdus, 78
Sandemanian Church, 28,
117, 122
Scattering of light, 23
Schönbein, C F, 125
Schroedinger, Erwin, 188
Scott, Sir Walter, 12
Select Committee of Houses
of Parliament, 48, 55
Selective adsorption of
solids, 23
Sensitive flame (of Tyndall),
156
Shakespeare, William, 3
Sheffield plate, 50
Sheffield, Lord, 13
Sherrington, Sir Charles, 190
Siemens, 43
Silvering, 50
Smith, Rev Sydney, 13
Smithsonian Institute, 211
Smyth, C P, 142-144, 189
Smyth, W H, 142, 151, 152
Solar neutrino, 26
Somerville, Mary, 46, 103
Southey, Robert, 12
Speke, John Hanning
(explorer), 189

Spence, Sir Basil, 188
Spiritualists, 125
Stanford, Leland, 167
Stokes, George Gabriel, 139
Stoney, Johnstone, 49
Sturgeon, William, 149, 150
Superionic conductors, 23, 87
Svedberg, T, 79, 94
Swansea Bay, 40

Table-turning (of the
    Spiritualists), 125, 127
Tate Gallery, 13
Taylor, Charles A, 141, 189,
    201, 202, 210
Telegraph, 39
Tenerife, 143
Tennyson, A, Lord, 136, 168
Tesla, Nikola, his Discourse,
    177-180, 190
Texas Monument, 99
The Times, 126, 127, 132
Thermistor action, 88
Thompson, Benjamin (Count
    Rumford), x, 4, 6, 19, 155
Thompson, Sylvanus P, 73
Thomson, J J, 49, 181-182
Thomson, William (Lord
    Kelvin), 3, 68, 75
Timothy, VI-10, 121
Transformer, 40
Trinity House, Corporation
    of, 3, 58, 131
Turner, J W M, 3, 116, 188
Tuscany, native caustic lime
    of (subject of Faraday's
    first scientific paper),
    24

Tyndall, John, 73, 75, 92, 117,
    123, 129, 153-155, 168,
    174-175, 211
Tyndall effect, 154

University College, London,
    38
University of London, Senate
    of, 60
Ur of the Chaldees, 183

Various Forces of Matter (M
    Faraday), 95
Vitalism, 27
Volta, Alessandro, 7, 19, 55,
    158
Voltaic pile, 55
Voltaic power, origin of,
    Faraday's views on, 157,
    159
Vulcanization of rubber, 23,
    38, 91

Wallace, Alfred Russel, 176
Watts, Isaac (*Improvement of
    the Mind*), 15
Wedgwood Museum, 26
Weinberg, Steven, 78
Wells, H G, his Discourse,
    176, 207-210
Wheatstone, Charles, 39, 72-
    73, 113, 139, 152
Whewell, William, 3, 46, 50,
    104
William IV, King, 56
Wilson, J Dover, 223
Wollaston, W H, 3, 13, 29-30,
    40, 159